PYTHONNET PROGRAMMING
Contextual GUI Logical Designs for Beginners

PYTHONNET PROGRAMMING
Contextual GUI Logical Designs for Beginners

Opubo G Benebo

Okumaye Publishing™
Alexandria, VA, USA

PYTHONNET PROGRAMMING
Contextual GUI Logical Designs for Beginners

Author: Opubo G Benebo

Published July 2021

Published in the United States
printing number
1 2021 First Edition

Book Catalog Information:

Benebo, Opubo G.
PythonNet Programming/Opubo G Benebo –1st Ed

Library of Congress Control Number: 2021914362

p. .cm
Including bibliographical references(p.) and index
ISBN-13 978-0-578-94310-7
ISBN-10 0-578-94310-7
1.pythonnet
2. pythonnet programming 3. pythonnet GUI paradigm 3.python paradigm I Title

Where to Send All Inquiries
All inquiries regarding this book may be sent by email to the attention of Chief Publishing Editor, cpe@okumaye.com

Okumaye Publishing Company, Inc.
Alexandria, VA 22314

www.okumaye.com

Acknowledgement 1 *The author expresses his gratitude to Enhance Knowledge LLC for the free access to some codes of in-active use software from which some of the codes have been acquired and presented even though some changes had been made to protect the confidentiality stipulations. The author is also grateful for the opportunity to develop the book for the use of the company staff and release to the general public such as might benefit from the contents.*

DISCLAIMERS

Preface

Python programming continues to be widely adopted, and is choice paradigm for AI and Data science and some scientific computations. Also, experientially, python programming is the most effective paradigm for supporting the inception of a Startup Venture. Software for most Business Activities is easily coded and readied for use, and also easy to continually upgrade, modify or integrated with another for expanded functionality. Such software, of course, need a user friendly GUI to support high productivity such as Startup Ventures are usually desperate to have at the early stages of founding.

GUIs are often imperative in some such applications obtained with python programming; and PythonNet Programming presents applications contextual GUI programming for fast-start. The book therefore is aimed at teaching the developing of pythonNet GUI as part of the development of python applications and for retrofit integration with existing python applications. Notably though, the codes presented in this book are not aimed at eliciting python programming or python codes optimization techniques; rather the objective, rather is to present the conceptual thinking guiding pythonNet codes development.

Further, this book proffers the rational Logical Designs Development approach in developing the insight into the interactivity between the components in the coding of pythonNet GUI software. Rational Logical Design programming is always best started off with a simple project and then built upon the same design until all the relevant features are implemented in course of the code development. This approach enables the learner to perceive of the changes that are made and the underlying rationale as the development work progressed; but even more significantly, the approach highlights the interaction between the various parts of the codes as such were being developed. Most importantly the training approach allows the learner to appreciate the various core aspects of the programming paradigm for the purposes of making reasoned coding.

The codes in this book have been developed with NetBeans 8.2 being chosen as such as the machine for the development was 32 bit. In that regard the NetBeans was used together with Cygwin which the NetBeans IDE recognized and integrated into upon start up. The downloading of python packages were performed mostly with the windows command application. The choice of Cygwin was particular instructive in that the codes were always placed either in a storage directory or flash-drive and so the execution of code required change of directory which was well supported by the Cygwin and not so much by the Windows Command interface,

The book is gem of expansive Fast -Start nuggets of example codes for some real immediate use for such contexts as dynamic data input into Listview, ComboBox, ListBox, Treeview and Directory Treeviews, and such hybrid objects as ComboBox with Treeview and mirror: Treeview with ComboBox; and use-application GUI as OpenFileDialog. Clear use of methods as invocation within custom functions and EventHandlers is lucidly explained, Effective choice of EventHandler

Finally, such unique applications of giving guided structured development of pythonNet GUIs for retrofit integration into Console Applications are proffered and the approach applied to several developments of current interest as Client Server systems such as Socket Client Server, gRPC Client Server for Scientific Computing, and of course the development of GUIs for Database Tables Design, MongoDB-aided Large Data Processing and generic Big Data Analytics.

The developer choosing to be instructed by the contents of the book should become an effective pythonNet GUI developer after concluding the study. Of course, not being a textbook, the developer will have to learn the contents by continually applying the materials of each chapter to the project in order to get a working knowledge of pythonNet programming. Happy reading!

Opubo Gbanaye Benebo

July 2021

Contents

Introduction

Developing GUI

The design of GUI in most Integrated Development Environments, IDEs, may be accomplished by static Design Method entailing the "Drag and Drop" of components onto a Form or programmatically. the former approach is used more often by beginners, for one thing there is visual appreciation of an action that is almost immediate, because the layout of the objects can be readily adjusted in design time. However, the latter approach is more challenging being more cerebral, because of the demand on the imagination of the layout at programming time, and to await test run for confirmation of the imagined appearance.

Yet for certain development conditions the programmatic approach is imperative and as such being familiar and comfortable with that approach is necessary for the performing of the required development task. In the presentations contained herein, the programmatic approach is taken because for one thing the learner gets exposed to and familiarized with the code development methods that associated with various functionality that otherwise is auto-performed at the design time. In order to reflect a strong parallel with the Static Design results, the development coding are aptly restricted to mostly Thin Clients. Indeed, the proffered coding in this book is strictly Thin Clients as it elicit very little with respect to Presentation Logic design.

The python jargon as used used has contextual meaning though all related and all different. For instance the software python is simply an interpreter that interprets codes written a specific format to the the Operating System of the computer. So then the term python is often used in referring to the Interpreter. Writing codes in compliance with a specific format for the interpretation by the Interpreter is then meant python programming

The PythonNet GUI Development

Developing a PythonNet GUI essentially requires extensive familiarity with the Windows development kits and objects which must necessarily be used freely. Many of these objects are available in online code samples and a developer can search for ones that may be missing here. In effect, the development here is not meant to provide a comprehensive exposure the windows GUI objects but rather as a tool for logical designs development.

C# as the guide with emphasis on the use of C# codes either verbatim or with translation. Significance of C# rudiments is good, and the translations for better coding Listview data entry is emphasized. In this regard the development of the concepts explicitly demonstrating the use of the standard Forms Properties, methods and Events, in Initialization Logical Designs in elicitation of the coding mindset well-suited for effective pythonNet GUI programming reflecting applications of pattern designs Clear use of methods as invocation within custom functions and EventHandlers is lucidly explained, Effective choice of EventHandler

Extending the approach to dynamic data imputation and extraction from the Window Forms objects, the instructive development is applied in in-use contexts for dynamic data input into Listview, ComboBox, ListBox, Treeview and Directory Treeviews, and such hybrid objects as ComboBox with Treeview and mirror: Treeview with ComboBox. These development shows the challenges of data manipulation consisting of data packaging and restructuring as with the ETL methods for subsequent input into these datasets-container Forms objects.

Invariably the object of a GUI development is often the development of an Application Main Window that bears the default set of menus of File, Edit, View, Tools or Configure and Help as characteristics of applications GUIs. Accordingly the coding for the Application Main Window is didactively presented in continuing the GUI development. The development here culminates with a template form code.

A complete development of the GUI is then evolved through the integration of judicious selection of GUI objects grouped onto a Form Dialog object, for processing data exchange involving a MongoDB document system implemented for use in the object Application.

Finally the development of GUI for and integration with console applications is vividly illustrated for ease of retrofitting console applications with GUI for simplified user need.Project oriented choices, Thin client logical design, Begin with the bare essentials and then show the build up for the thinking guiding Thin Clients for such unique Client-Server applications as Socket Client-Server, and gRPC Client-Server Computing

Chapter 1

Basic PythonNet GUI

1.1 GUI Developing Process

Developing Graphical User Interface, GUI, whether or not a pythonNet GUI for an application is in essence an incremental customization of the basic GUI, essentially a logical design building up of the basic GUI. For the purposes of a more reasoned design of GUI, a programmer needs to be continuously cognizant of the context of design as a case of continual customization. Such awareness is significant to guiding the developer in developing effective codes. Interestingly, this is not obvious to some GUI programmers who simply instinctually perform the coding without actually thinking about it.

A good starting point then is the careful and structured eliciting of this awareness in the coding of pythonNet GUIs.

1.2 The Basic Interface

The bare essentials graphical user interface, GUI, consists of no more than the reproduction of the Windows Forms from which the GUI as being derived followed with specialization and and slightly customization for the display screen, otherwise nothing else as is shown in Fig 1.1. However, the Basic Interface is then

Figure 1.1: A Basic GUI of a PythonNet Application

usually further specialized depending on the use-need to support menus, toolbars and statusbars as the developer may prefer. Of course, the actual operations for which the GUI is being developed, are implemented in the space bounded by the walls. The consideration of the these anticipated specializations are instructive in understanding in the development of codes.

1.3 Menus Coded Basic GUI

The specializations of the code to support menus is generally enabled with the inclusion of these codes for the creation of menustrip, resulting in the emplacement of a strip of space onto which the menus may then be embedded as with the use of the code for, say, File, the end result of which is the modified forms of the GUI as shown in Fig.1.2 in which the file menu is displayed. Of course, the developer may

Figure 1.2: A Basic pythonNet GUI Specializedfor menus

add as many menus as the application design specification stipulates or demands.. The stand set of menus deemed minimal for every every application, is discussed in Chapter 8

1.4 Specialization for Toolbars

Then, of course, certain menus and sub-menus that are used frequently or considered to be used quite frequently are often bundled into Toolbars. Therefore Toolbars therefore are not necessary but are convenient and so advised to be included in the development process. As with the menu, each item of the toolbar is added to the toolstrip as per the use need of the application. The inclusion of

Toolbar is as with the menus based on a sort of toolbar strip, as shown in Fig.1.3 with a lighter color strip at the top. The menu strip is sightly more graish while the toolbar strip is white. Further, the presence of the toolbar strip for containing

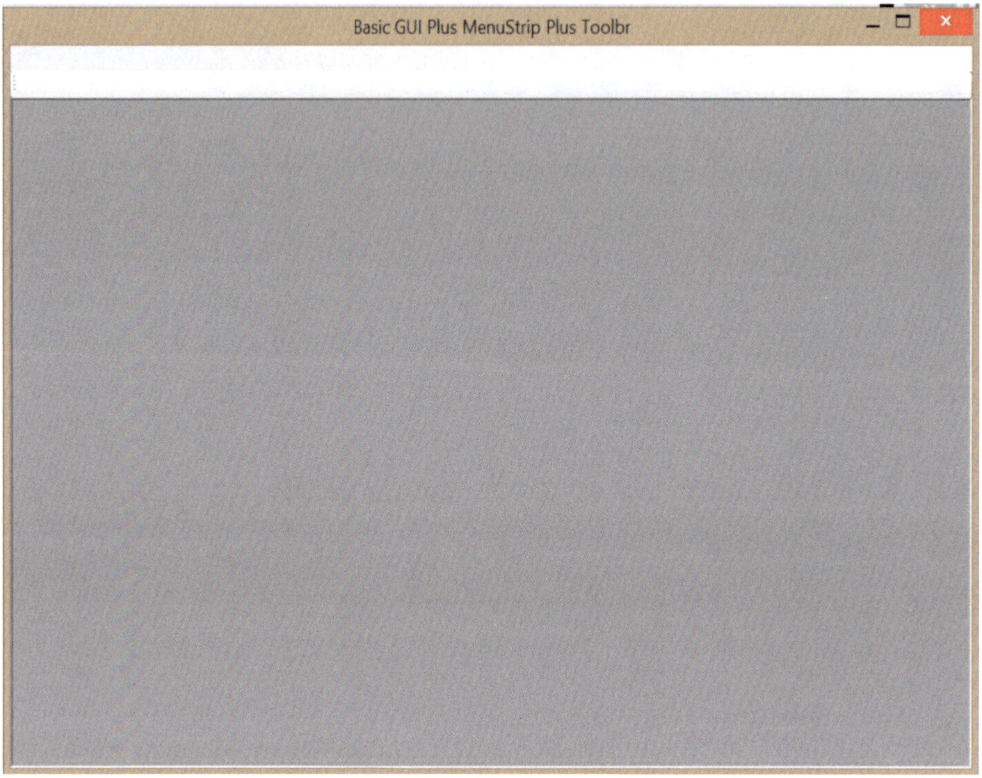

Figure 1.3: A Basic pythonNet GUI plus menu and toolbar

the tools is reflected with the wider strip. More in depth presentation will be made later but suffice it that the items of the toolbar can be represented with images.

1.5 Specialization for StatusBars

Very often being able to observe the progression of the tasks being performed is desirable and such is usually proffered with the inclusion of the Statusbar in the Basic GUI. The inclusion of the StatusBar at a minimum modifies the Basic Interface into the form shown as in Fig. 1.4 with yet another strip added to the Basic Interface, but albeit at the bottom bounding wall. The StatusBar is usually

Figure 1.4: A Basic pythonNet GUI plus menu and toolbar, and statusbar

placed at the base of the GUI as a sort of ruler strip on which is displayed the tasks as such are being performed.

Chapter 2

Getting Setup

2.1 The Development Environment

The coding and execution of the code towards the development of the GUI occurs within a set collection of functional software components which constitute the Development Environment. then, of course, such environment are of myriad variation as there are developers.

The variations of the environment comes about from the collection of tools that each developer chooses to use for performing the coding and code execution.

2.2 Development Tools

The collection of development tools can consists of some combination of any of these listing:

- Source Code Editor
- Integrated Development Environment, IDE
- Cygwin
- GitBash

- Python Software
- Python.net Package
- vs_BuildTools

Without regard to the order of the listing of the tools for setting up the development Environment, the first and foremost tool to implement is the deployment of the python system to the end of installing pythonNet development capability. So then first installing the system is foremost and is presented in the order.

2.3 Source Code Editors

At a minimum, a source code editor is required and must be installed. The Source Code Editor, however, may be substituted with a Text Editor, though the Text Editor often does not provide text highlighting and other features of convenience as found in Source Code Editors. Notably, Source Code Editors have to be used in conjunction with terminal by which to access the compiler, both as separate software components, that as such must also be installed.

Several Source Code Editors are available from which a developer may choose one or as many as the development task demands and according to the interest of the developer. The most common Text Editors starting with the most elementary to the complex ones may be listed as Notepad, Textpad, Vim. EMACS, VS Code, Atom to mention a few. Notepad and Textpad generally are just mere text editors with any frills to them; yet even between the two, Textpad has more features better suited to code development.

2.4 Integrated Development Environments, IDEs

Generally Integrated Development Environment is an application software that often provides an environment consisting of a source code editor, a compiler, debugger, and any other tools that the vendor of the apllications deems essential for the development style being projected. These tools have been integrated to give a seamless use by the developer with minimal of reconfiguration to have a working software. An example of one such software is the NetBeansIDE, Visual Studio.

There are also specialized IDEs which come prepackaged with development kits consisting of components specific to the use context as designed for the user-developer. These prepackaged development environments often are platforms being designed for specific development type endeavors. Among this group are Anaconda, Spider, PyCharm and Django to mention are few. Though so packaged, some still do not have Text Editors and must yet be interfaced with Text Editors, one such platform is the Anaconda platform that use Vs code. The choice of an IDE of the platform category, however, often precludes the need for the python software as the IDE usually includes the python software as an integrated component. The developer generally chooses to use according to need and taste. Each of these, of course, has its advantages and disadvantages.

2.5 Command Interfaces Tools

While the tools of Cygwin and GitBash are specific, the IDE and text Editor are varied and the choice depends on the preferences and idiosyncrasies of the developer.

2.6 Python Software

The installation of the python can be achieved by the simply installing the software as downloaded from the python.org site as listed above, or by the installation of one of the several python platforms, or IDEs. So then the inclusion of the python software very often is a part of the process of making Text Editors useful as tools.

The python interpreter generically called python software is developed by the Python organization and can be downloaded for deployment from the organizations website:

`https://www.python.org/downloads/`

and once at this site, the selection of the operating system of the development machine is made and the download begun. Upon completion of the download, the developer simply double clicks on the download and the follow the instructions to complete the installation.

The successful installation can be verified by going to the user directory and searching for the python folder. In the modern Windows Systems, it can be found usually at the following default directory:

```
\Users\yourname\AppData\Local\Programs\Python\Python38-32
```

while in Linux and UNIX type systems it could be found at the default directory :

```
\Users\yourname\AppData\Local\Programs\Python\Python38-32
```

The working condition of the installation can then be verified by running certain commands at the command lines, for instance to ascertain that version of the python installed the command

```
\$ python version
```

After the completion of the validation of the installation of the python software, then the development system is ready for the installation of pythonNet on it.

Although the process of downloading and installing a python software on the computer machine, a developer may sometimes want to use several versions of the python software installed on the machine

2.7 Installing Python Packages

A driver of the popularity of the python programming language is that as an Open Source, there are available many packages developed over the years by the community and are made available for developers. Admittedly other languages also have packages of various forms but with such languages as C and C++, the developer actually has to develop the application specific packages in the form of

DLL (.dll) or as OS(.os) for subsequent inclusion in all future development project tasks. So a significant advantage of python over some other languages with respect to the ability of a developer to develop python programs is the absence of the need for the developer to develop each of every one of the application-specific packages.

Having the application specific packages for inclusion into GUI development as such is essential in getting the efficacious development environment.

2.7.1 Installing pip

These packages are often installed using a package installer, termed pip, and which must be installed following the installation of the most currently stable version of the python interpreter as above.

The software pip is developed by an open-source organization and can be downloaded from the website

```
https://pip.pypa.io
```

however, since python 3.0+ a copy of pip has always been co-installed with the python interpreter and is usually found in the python home directory at the site-packages. So if the pip folder in the python home has not be tampered with then the pip can be installed by performing the following steps:

1. Change the directory to the python home directory: For Windows OS computer the folder may be as given

   ```
   C:\> cd C:\.....
   ```

   while for the UNIX-like system as given

   ```
   C:\> cd C:\.....
   ```

   after the successful change of directory as confirmed by the response as

   ```
   C:\.....>
   ```

2. Then the pip is then installed with the command

```
C:\.....> python install pip
```

and this should install the copy of the pip in the home folder of python. Often this is obsolete as of the time of installation of the python interpreter and there will come up suggestion to upgrade to the newest version of pip as shown Fig. 2.1

Figure 2.1: pip upgrade advisory

and the pip should invariably be upgraded as advised as per the command:

```
C:\> python -m pip install --upgrade pip
```

which should begin the process of installing the newer version following the removal of the obsolete version as may be observed in progress in Fig.2.2. The end

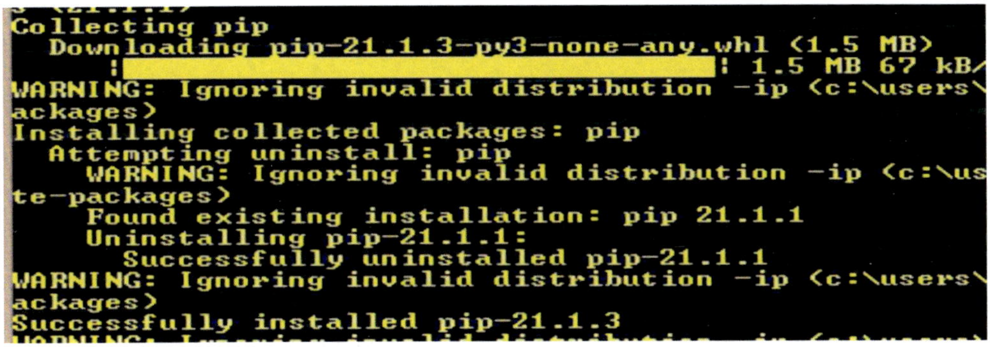

Figure 2.2: pip successfully installed notification

result of it all is that the most current version of pip will have been installed and work can progress.

2.7.2 Context Python Packages

The specific collection of packages installed, however, depends on the needs and objectives of the developer. So for pythonNet programs there are also packages that are required to be installed for the needs of the developer.

Although each developer will install as preferred, the packages installed with which the contents herein are developed include the following list

- numpy
- fmatch, re
- openCV

and all these packages are suggested to be installed at the very minimum. Installing each of the packages is accomplished with the command:

```
C:\> pip install package-name
```

and then hit Enter and the package gets installed in most cases, although every so often there may be some issues.

2.7.3 Installing PythonNet Package

As with the python software the pythonnet software is developed yet by an organization, and is available at the website:

`https://pypi.org/project/pythonnet/`

as python package in the form Python.Net. Again having navigated to the correct site, the software is downloaded as an executable and the installed by double clicking the executable.

2.8 Code Developing Essentials

Now that the python development environment is deemed set up, some codes are written to validate the functionality of the set up of the environment.

Starting off, the simple renown one line program of "Hello, World" is written at the python interpreter alalso called the python shell. For that, start off command interface of the system and at he command prompt, type in as shown

```
C:\> python3.x
```

where the x is the minor version number of your python interpreter, and hit Enter, and the response should show

```
>>>
```

Now the interactive tests can proceed earnestly, and so type in as desired the greetings:

```
>>> print(f"Hello, World"')
```

and again hit Enter, and and the response should show the greetings as in

```
Hello, World
>>>
```

The same could be done with GitBash: start of GitBash and at the command prompt , type as shown

```
$ python print(f"Hello, World"')
```

and hit Enter, and as with the Windows case, the response should show the greetings as in

```
Hello, World
$
```

Clearly there is not a difference between the use of the Windows Interface from the GitBash Interface but there will come about a difference for which the two interfaces are being discussed.

This next step could also be performed with an IDE, but it is much more insightful to use a Text Editor; so open up a text medium, say, Notepad or Textpad or any other text editor of preference. Copy the simple command

```
print(f"Hello, World"')
```

and paste on the open text document, and save as, say, "hello.py" in folder in the Local Directory, often labeled "Drive C", and then once again open up either command interface and type in as shown

```
$ python ./hello.py
```

and hit Enter:, and again as before the response should show the greetings as in

```
Hello, World
$
```

First and foremost, worth clearing some observations. In the above code of "./hello.py" the prefix "./" to the file name, means, in current directory. Its introduction at this stage is apt as it will become useful soon enough and more so when it is required that Directory Change be made to a specific Directory.

Just in case more python programs may be created in the future , then to keep all in one place, next a directory is created and named "hello" still on the "Drive C", and then the hello.py file is moved into this folder. Once again the object would be to able to execute the simple program. For that again the command interface is opened up and the following typed in as shown and then once again hit Enter:

```
c:\>python .\hello\hello.py
```

and once again the response from the system shall be as follows

```
Hello, World
c:\>
```

While all this is well and good, and that a folder has been created to hold the files, one can proceed to start coding; however, because the active drive is often attacked and may have to be wiped off and the system restored it is good practice to set the code in the storage Drive, say, Drive E.

Again the code could executed as before like so

```
c:\>python E:\hello\hello.py
```

and once again the response from the system shall be as follows

```
Hello, World
c:\>
```

A good observation is that at this stage Drive E in the form of its identifying letter E has to be included in the file-path of the python program. Although it is much trouble to key in the filepath at this time, there are cases when the filepath can be so long that repeated typing of the full filepath just to execute the code becomes burdensome, and a better approach is usually preferred. One approach is to make a Change of Directory as to execute the code without having to include the directory path in the file path.

However, in some cases the Windows Command Interface would not do anything if the command as written is executed:

```
c:\>cd E:\hello
```

The code would remain static here and fail to change the directory. Yet a GitBash command interface would do this:

```
$ cd E:/hello/hello.py
```

and the result would be

```
$/E:/hello/hello.py>
```

meaning that the directory has changed and this is the advantage of having a different command interface instead of that of standard Windows command interface. Moreover, the python program could once again be executed with the simple command:

```
$ python ./hello.py
```

and again as before the response should show the greetings as in

```
Hello, World
$
```

Clearly the use of the fileprefix, "./" in further designation of the directory location of the file becomes obvious.

Another test of the folder is the use of functions execution to check folder accessibility and pythob module import capability. For that the "hello.py" file contents are put into a function definition as in python:

```
def Greetings():
        print(r"Hello World?")
        return True


if __name__=='__main__':
    Greetings()
```

Again this file is emplaced in a folder named "greetings" but this time saved as "phello.py". In the another folder "hello" is placed yet another file, on which are written the two lines of codes:

```
import greetings.phello as phello

phello.Greetings()
```

and is saved as before with the file name "greetings.py". Next ,the usual change of directory is set to the "hello" folder, and yet again the filename invoked like so:

```
C:> python ./greetings.py
```

Of course, this should not work generally. For this to work, the folder containing the program file must allow the import directive, and for that a file named ini.py must be included in the folder named "greetings". The init.py file has written on it at a minimum, "__init__" as all the code. The lines before and after the word, "init " are double underscore, of course. A repeat of the steps above should result in the execution of the function code.

So then for every folder from which a python file needs be imported into another file, then an "__init__" bearing file must be included in the folder. It is also instructive to actually have the two folders: greetings and hello form a directory of nested folders such that the file-folders directory tree has the form:

```
hello
    hello.py
    greetings
        greetings.py
        init.py
```

where the files have the .py file extensions and the folders do not. Nested codes file folders is very useful for file management when developing large scale python software, as at such times codes are usually very long and dividing the codes into several files and then grouping related files into specific folders enables manage the code development effectively.

2.9 Running the Code

So far, for simple code tests, the code has been shown to be testable with an execution invoke: A code written on a file has been the filename as, say, "mypynet-program", the extension .py applied, then using any Command interface a python invocation made to execute the code by typing in at the command line –if using GitBash then type:

```
C:> python .\mypynetprogram.py
```

and hit Enter; or if using Powershell of Windows (not the regular Command Exec0 and then type in

```
PS python ./mypynetprogram.py
```

and hit Enter. the result of this process has been to provide the result in the sample cases of greetings of hello. The option of executing the code either with the Windows Command App or the GitBash Command brings to question, which is preferred for the purposes of developing python software. The choice of tools, of course, is quite varied.

However, the code is run during the development by the method because of the python execute function:

```
if __name__ =="__main__":
    Greetings()
```

but without the function included in the module and so making the module an executable module, an alternative approach has to be used and one of such approaches is given by C# Embedded Compilation

2.9.1 C# Embedded Compilation

During development the file bearing the above code is stored in the same development folder and stored as applicationName.csc, In the approach of using C# for developing the code, the use of Py.GIL is adopted and the C# code to use is given as

```
using System;
using Python.Runtime;

public class Test
{
    public static void Main()
    {
        PythonEngine.Initialize();
        PythonEngine.BeginAllowThreads();

        using(Py.GIL())
```

```
            {
                try
                    {
                        dynamic module string[] arr = argv.ToArray();
                        Runtime.PySys_SetArgvEx(arr.Length, arr, 0);
                        Runtime.CheckExceptionOccurred();
                        arguments = Py_BuildValue("string", [])
                        Py.Import("module");
                        dynamic result = module.function("argument");

                        // Console.WriteLine(result);
                    }
                    catch(Excepiton ex)
                    {
                        // Handled Exception
                    }
            }
        }
    }
```

where the module is the fileName, "C:

...

greetings.py"'and the function of the module is the function Greetings() such that the dynamic execution step is actually as

```
dynamic result = greetings.Greetings( null);
```

and is repeatedly executed as follows

```
PS :\> CSC ./applicationName.csc
```

Yet the invocation of this script is the same as for the earlier scripts as explained above.

2.10 Application Compilation

However, a necessary object of most development is the deployment of the development on machines other than the machine on which the development was carried. in effect there is the need to make the codes available to user on the users machine. For that object, the code must necessarily be compiled either into bytecodes or machine codes.

The compilation can be accomplished in several ways, most of them compiling into a console application - sort of. Some option of compilation include:

- pyinstaller Compilation
- C# embedd compilation
- Bytecode compilation

In order to see the UI of Fig. ?? the code is simply typed into a Text application, say, Notepad or Textpad and the saved into a file that is given the document extension, .py, and then compiled by one of the methods listed above and delineated presently.

2.10.1 pyinstaller Compilation

A python package, pyinstaller, compiles the pythonNet codes that develop GUI into an executable,.exe, that launches with a double (or as the case may be, a single) click on the icon of the output. as with most python packages, the pyinstaller can be installed using pip as with this invocation:

```
PS :\> pip install pyinstaller
```

Following the installation, he developer can compile the application using the simple invocation:

```
PS:\> pyinstaller --onefile applicatioName
```

where the applicationName is the name of the main file of the application. The "–onefile" command line option makes the code distributable to other machines. However, the Command Line options of pyinstaller are many and diverse and the the –onefile though functional is not one of the options. A developer would effectively become conversant with the features offered by the options and make the appropriate choices of options as suites the goals for the codes development. At the end of the compilation, the output is

```
Compilation is successfully completed
```

When the folder where the application files are placed is accessed, there will not be included a sub-folder, dist, which stands for distribution, and in which will be found a file, applicationName.exe, which is the executable application that could be executed in any other computer that has python interpreter installed.

Particularly interesting with the use of pyinstaller to creating the applocationName.exe is that the application file will either have to outright invoke the name of the primary function, say, Greetings() as the last line of the file otherwise include the python code

```
if __name__ =="__main__":
    Greetings()
```

else the application can not start. However, both tasks for enabling the starting up of the application also first creating a Console and before starting off the non-Console GUI application. An obvious preference would be the non inclusion of the start of the Console. For this preference the invocation from C# yet again offers an option.

2.10.2 C# Embedded Compilation

Indeed, the same scripts as before but with some changes in parameter values is all it takes to accomplish the objects of the compilation. In both cases however, the pyinstaller is used to generate an exe of an application file that does not invoke the primary function either directly or by the use of the python invocation method.

So then the objective is to invoke the execution of the applicationNae.exe through the C# scripts, which is now modified as follows:

```
Using System
Using System.Diagnostic
Using System.IO
Using System.Threading.Tasks;
Using System.Windows.Forms;

private void run_cmd(string cmd, string args)
{
    ProcessStartInfo start = new ProcessStartInfo();
    start.FileName = cmd;
    start.Arguments = args;
    start.UseShellExecute = true;
    start.RedirectStandardOutput = false;
    using(Process process = Process.Start(start))
    {
        using(StreamReader reader = process.StandardOutput)
        {
            string result = reader.ReadToEnd();
            Console.Write(result);
        }
    }
}
```

and the cmd variable in the C# codes is actually the full path to the python.exe, and the args is the application full filename, greetings.py and cmd line args; and with two changes, essentially a switch: The assignment to the parameter: start.UseShellExecute; is changed to true, and the assignment to the parameter:

start.RedirectStandardOutput; is changed to false. Hence two situations obtains, the system locates the python.exe folder and loads it for use, and the results are no longer written to the outcome except where explicitly demanded by the developer.

Also the value of the assignment cmd is changed to simply python.exe, and the args is changed to applicationName.exe. Finally the C# script is the compiled with the C# compiler, Robyn, as with the command at compilation

```
PS:\> Robyn ./applicationName.csc
```

to generate a new .exe file in the same folder.

2.10.3 Bytecodes Compilation

Compilation of the pythonNet codes into bytecodes offers another approach to generating a standalone application.

However, because the bytecode is not fully an exe of the application, and requires explicit access to the python interpreter functionality, its execution is better addressed with the C# scripts used for the development as in Section 2.9.1. Effectively, though not suited for use with the execution, .exe file, however it is still useable in executing bytecodes used in the context as

```
using System;
using Python.Runtime;

public class Test
{
    public static void Main()
    {
        PythonEngine.Initialize();
        PythonEngine.BeginAllowThreads();

        using(Py.GIL())
```

```
            {
                try
                    {
                        dynamic module string[] arr = argv.ToArray();
                        Runtime.PySys_SetArgvEx(arr.Length, arr, 0);
                        Runtime.CheckExceptionOccurred();
                        arguments = Py_BuildValue("string", [])
                        Py.Import("module");
                        dynamic result = module.function("argument");

                        // Console.WriteLine(result);
                    }
                    catch(Excepiton ex)
                    {
                        // Handled Exception
                    }
            }
        }
    }
```

but with the bytecodes being the modules, and the arguments passed as command line options.

However as will become obvious the function directly executed when the application is a pythonNet application is given by

```
def GUI_main()
    Application.EnableVisualStyles()
    app = Greeting()
    Application.Run(app)
    return True
```

and the application code that is compiled to a .exe is without the python invocation:

```
if __name__ =="__main__":
    main()
```

or the explicit invocation of the function as by

```
main()
```

because of the GUI codes execution in an infinite loop define through pyGIL that keeps open the GUI for as long as the end-user keeps it so for use or otherwise.

Chapter 3

PythonNet DataTypes

3.1 Typed and Non-Typed Paradigms

Programming languages are often characterized as Typed or non-Typed, and the coding is developed accordingly. The typing of variable enables the storage mechanism to allocate the correct memory space for the storage of the variable.

Generally typed languages have all variable declaration preceded with the type of the data as per: [1]

```
datatype variable
```

On the other hand, non-typed programming languages do not require the assertion of the datatype of the variable. The declaration is simply made as

```
variable
```

rather the type is made by the compiling or interpreter mechanism based on the assignment to the variable, a s result most non-typed languages require that all variable declaration have the assignments be made at the point of declaration as so

```
variable = assignment-type
```

and when the assignment type is string then the compiler of interpreter dynamically types the variable as of the string; and if integer then the variable typed as integer, and so on.

When developing pythonNet codes, two sets of Data-Types have to be considered by the developer. First there are the Data-Types of the python paradigm and then there are the Data types of pythonNet paradigm which are derived Data-Types of the dotNet (.Net) platform.

Because the .Net Platform usually requires the specification of Data-Types, and python does not, the definition and assignments of values to variables must be cognizant of the demand for the Data-Type for compatibility to obtain. An illustration of the issue is best proffered by this assignment to variable:

```
self.readerContainer.Size = WinsDraw.Size(int(self.Width/4 + 10),
                                          int(self.Height))
```

In this size assignment to *readercontainer* of size value the .Net Platform function of Size requires the values in the parenthesis be integer numbers, and such may often be used, however, there are cases as in the above when the values are desired to be pegged to base Form size height and width; and those cases the values must be cast into the specific Data-Type which in this case happens to be the *int* type.

3.2 Python Data-Types

The Data types of python that the Interpreter uses as reference in dynamically typing all variables declarations are the following listed types, delineated presently

unicode

int

long (Py 2.x), int (Py 3.x)

bool

float

list (sequence protocol)

None

object

decimal

dictionary

tuple

The definition and usage of these Data-Tytpes are found in just about every python programming source material[2;3] and so are known to the python programmer developer. However, the Data Structures of List, Dictionary and Tuple are extensively presented in the later Chapter 7 when their uses in Data Flow are presented.

3.3 The .Net Platform Data-Types

The Data types of .Net, dotNet, that the Interpreter uses as reference in dynamically typing all variables declarations are the following listed types, delineated presently

string, char

Int16, UInt16, Int32, UInt32, Byte, SByte

Int64, UInt64, UInt32

Boolean

Single, Double

IEnumerable

null

Decimal

Struct

array

The definition and usage of these Data-Tytpes are found in just about every .Net Platform source materials[4;5] and so are readily accessible but tangential to the presentation of the pyNet programming as presented.

3.4 Type Maps

The coding of the .Net API within the python program requires that the Data-Types of the API be represented by the equivalent python Data-Types in order to ensure compatibility. For that need, the two distinct Data-Types needs to be mapped to each other to elicit equivalencies in datatype. From the comparison of the delineations of each of the Data-Types of the two categories, an equivalency map could be developed of the associated given :

.NET Type	Python Type
String, Char	unicode
Int16, UInt16, Int32, UInt32, Byte, SByte	int
Int64, UInt64, UInt32	long (Py 2.x), int (Py 3.x)
Boolean	bool
Single, Double	float
IEnumerable	list (sequence protocol)
null	None
Decimal	object
Struct	dictionary
array	array, tuple

By this association, then, a developer only needs to use the association as a Look-up for selecting the correct equivalent data-type for use in the codes.

Chapter 4

PythonNet GUI Basics

4.1 PythonNet GUI

The GUI developed with pythonNet entails the use of Windows Forms, as base objects and a collection of nested objects related to the Form as nested objects instantiated from Nested Classes as defined in Object Oriented programming, OOP, derivative the concept as originally developed with the C++ paradigm. Effective development of GUIs with pythonNet therefore requires some understanding of the OOP paradigm and the concept of Classes and the Nested Class structures. OOP

4.2 Classes, Members and Methods

Strictly, a Class is description of a category of existence or life which description is a generic characterization of a whole collection of objects of that category whether tangible or intangible[1]. A classic example is given by the generic noun, Box, which as implied, generically characterizes Trunk Boxes, carton Boxes, Luggage [types], portmanteau, and what-have-you. so then these varied forms of six-side surfaces containers can be described programmatically as of the Class Box. The basic programmatic description of the category of Box may be written in the paradigm of C++ as: ,

```
class Box:
{
    members
    Box();

    Methods()

    ~Box()
}
```

The two functions: Box() and Box(); that bear the name of the class are respectively termed Constructor and Destructor, and respectively have the task of instantiating and destructing an object of the class in-the memory. Effectively the object is created with the invocation of the Constructor function and destroyed with the invocation of the Destructor function.

Members are generally defined by assignment of values to each of them during the class design or during instantiation by making such assignment within the Constructor function: Constructor(); which effectively designs the object at instantiation. The assignments are made as simply as say, for Length and Width:

```
length = 16
width = 4
```

4.3 Nested Classes

A very interesting feature of Classes as per the design definition is that these constructs can also contain similar constructs. The Outer class become containers for the inner classes as presented:

```
class OuterClass:
{
```

```
        properties:
        InnerBox smallbox:

        Box();

        Methods();

        ~Box();

        class InnerClass:
        {
            nnerBox-Properties
            InnerBox():

            InnerBox-Methods()

            ~InnerBox()
        }
    }
```

Although seemingly simplistic, this feature of classes is very effective in the description of complex objects and is particularly useful in the coding of engineering systems and complex biological systems. Of course, the container properties is more often used in the development of tools for software than the significance in describing the integration of objects.

In any case, of interest in the efforts of this book is the usefulness derivative of the container quality than any other; and it is that feature as used in the development of pythonNet GUI that is of consideration. Indeed as would be observed in course of the development of various GUI objects, additional lasses are nested either in the main window or in the GUI Dialog design specification class itself, effectively involving the use of Nested Classes.

4.4 Developing with Windows Forms

Window Forms is in essence a class with numerous inner classes all of which constitute the objects with which .Net GUI are developed. These objects include the usual visible objects of

- Label
- TextBox
- RichTextBox
- Button
- ProgressBar
- ComboBox
- ListBox
- ListView
- Panel
- GroupBox
- Container
- TabbedDialog

and very many others that support the instantiation of these objects. Conceptually therefore the Windows Forms object may be viewed as instance of a Form class of the description:

```
class Forms
{
    properties:
    Forms()
```

```
        Methods()

        ~Forms()

        class TextBox:
        {
            extBox-Properties
            TextBox()

            TextBox-Methods()

            ~TextBox()
        }
        .
        .
        .
    }
```

where the vertical ellipsis represent the rest of the other objects as listed above.

Of particular importance is the implicit functional groupings of these Form Objects, and the consequently the requirement in the design process that these functional grouping features demand. Specifically, these functional groupings obtain: Data Display and Entry Functionality, Data Container Functionality, and Objects Container Functionality. The Data Display and Entry Functionality group consists of these objects:

- Label
- TextBox
- RichTextBox
- Button

- ProgressBar

but require no specific requirements of data manipulation for any of each object. The Data Container Functionality group consists of these objects:

- ComboBox
- ListBox
- ListView

and require special additional Properties, Methods and Events to enable the manipulation of the data contained therein. Undoubtedly cognizance with these requirements are necessary for developing good GUI applications. The Objects Container Functionality group consists of these objects:

- Panel
- GroupBox
- Container
- TabbedDialog

and also require special additional Properties, Methods and Events for manipulating the objects - in terms of visibility and accessibility; that are contained in each container,

Expectantly, some of these other inner class may be used in course of the presentation and will be pointed out as such at such times of use. However, suffice it that these objects have many properties that are common to all the classes, then, of course, there are also class-specific properties. Similarly there are also methods and events that are common to many of the objects and then there are also the class-specific methods and events. each of these are presented for consideration for uses based on the understanding of their uses.

4.4.1 Windows Forms Common Properties

As observed earlier by virtue of the derived character of the objects, these objects have the same properties as the base class and as such have certain set of properties that are common across all the objects. The list of these common properties are provided in Appendix-A-properties; however, the listing is not presented as exhaustive, rather as representative of the commonly used properties. even so not all the listed properties are regularly defined in the development of GUI.

Obviously then, the properties of each object are composed of the common properties and object-specific properties. So for each object certain common properties are define and must be defined for most appearances. The most commonly used properties in the design of Form Objects are the ones as listed:

Name

Text

Location

Width

Height

Point

Size

Alignment

that help design the appearance of the target objects. Every GUI design coding necessarily effective includes these properties and also assigns specification to each one.

4.4.2 Windows Forms Common Methods

[6]

As observed earlier by virtue of the derived character of the objects, these objects have the same Methods as the base class and as such have certain set of Methods that are common across all the objects. Some of these common Methods are as provided in Appendix-B-Methods; however, the listing is not presented as exhaustive, rather as representative of the commonly used properties. even so not all the listed methods are regularly defined in the development of GUI.

Obviously then, the properties each object are composed of the common properties and object-specific properties. So for each object certain common properties are and must defined - and most appearances.The most commonly used properties in the design of Form Objects are the ones that help design the appearance of the target object.

new ClassType()

Refresh()

BeginUpdate()

EndUpdate()

Dispose()

These functions are often written within another developer-defined functions to achieve certain objectives. For instance the function Dispose() culd be used inside a destructor function, say as with the Box class, like so:

```
~Box()
{
    Dispose();
}
```

Of course, this definition would ordinarily be written inside the class definition otherwise a Binding would be required leading to the construct of like so:

```
Box::~Box()
{
    Dispose();
}
```

as usually the case when coded in the Implementation file for data hiding.

4.4.3 Windows Forms Common Events

As observed earlier by virtue of the derived character of the objects, these objects have the same events as the base class and as such have certain set of events that are common across all the objects. The list of these common properties are provided in Appendix-C-Events; however, the listing is not presented as exhaustive, rather as representative of the commonly used events. even so not all the listed events are regularly defined in the development of GUI.

Obviously then, the events of each object are composed of the common events and object-specific events. So for each object certain common events are required and must be defined.The most commonly coded Events in the design of Form Objects are the ones that enable the developer react to the most frequent interaction of the user:

OnClick

GetFocus

OnGetFocus

OnLosefocus

OnKeyPress

with the target object, which require a binding of the object to the interaction being observed for reaction, so say the developer wishes to respond to the user clicking the object. So the developer necessarily forces the object to raise a specific Event by associating an EventHandler to the Event by one of two formats as:

```
obj.Event += obj.EventHandler
```

or by explicitly relating the Event raised to the System Component Model as

```
obj.Event += System.ComponentModel.CancelEventHandler(
                                            obj.EventHandler)
```

Then the developer may code EventHandler function to cause to obtain the preferred outcome of the developer.

```
Form.EventHandler(obj, sender, eventargs):
{
    if sender == EventRaiserObj:
        preferred outcome generated
    return True
}
```

This function will execute when the label object is clicked will copy the Title of the label object and assign it to the variable textmat which the developer may the used as seen fit . In yet a trivial but demonstration effective form the developer could also be free to code like so:

```
Form.Click()
{
    if sender == EventRaiserObj:
        obj.Dispose()
    return True
}
```

The significance to the codes is that the developer uses the EventHandler to effectuate certain behaviour of the object as desired by the developer of the specific object.

4.5 PythonNet C#

Python .Net is as explained in Chapter a development based on C# and so most codes from c# can be directly used with minimal modifications particularly when dealing with properties design.

By and large most coding of the assignments to the properties of the objects as provided in the Microsoft C# programming language carried over directly into the pythonNet programs. As such the programming in pythonNet can benefits significantly with some understanding and skill-set of C# Programming, yet a pythonNet GUI developer can still effectively develop efficacious application specific GUIs. Effectively a pythonNet develop is encouraged to review codes in C# for inclusion in pythonNet though such inclusion should start with a test of useability.

4.6 From C# To PythonNet

The coding of pythonNet GUI reduces in essence to the effective translation of the corresponding c# codes to the extent that the developer has access to such codes. Obviously for the common properties the assignment coding of the C# is the same and as such emplaced in the pythonet code verbatim, as illustrated: Snippet of C# code as known or found in public domain:

```
this.Length = 16
this.Width = 4
```

would translate verbatim into a pythonet codes as follows:

```
self.Length = 16
self.Width = 4
```

Except for the replacement of the "this" in the C# paradigm with the "self" in the pythonnet paradigm, the assignments are exactly the same. So then for properties the assignments are exactly the same but for Object-type casting for compatibility.

The translation of the C# Constructors into pythnNet Constructors and the coding of Destructors in pythonnet rather require some care and attention while the C# approach of constructor invocation takes from the C++ pointer style and uses the "new" command the pythonNet format being derivative of the python style uses C++ reference style and actually self references: First the C# Constructor:

```
\#this is c\# code
ClassType obj = new ClassType()
```

Invocation of c# object instantiation upon translation into pythonNet is replaced with the non-pointer C++ style invocation

```
\#this is pythonnet code
obj = Windows.Forms.ClassType()
```

The Constructor translation into pythonNet codes is is given as "__init__(self):" however, this function is actually usually coded as a hydrid of the properties and the Constructor proper, and as a result it more completely coded as:

```
def __init__(self):
    properties
    InitializeClass(self):
```

The Destructors as with the Constructor does not bear the name of the class either but rather is titled "__del__ " as del for delete, and is coded as

```
det __del__(self):
    pass
```

Coding the Destructor function for the pythonNet may choose, however, to be more resources conscious and always force the objects going out of scope to release all resources utilized in the instantiation of the object with the invocation of the Dispose method, as per

```
def __del__(self):
    if not isDispose:
        Dispose()
```

This form of coding though somewhat of more work should be memory efficient and allow for more efficient use of resources in programs that involve the use of large quantity of resources, and is encouraged.

Combining all these to construct a guidance for the translation of a typical C# into a corresponding pythonNet code, the general construct of a typical C# code is instructive and has the structure as per:

```
using system
namespace pynetspace{
class Box{
    nt length = 16;
    int Width = 4

    public:
        Box()

        ~Box()
    }
}
```

The translation of this class into a pythonNet class entails the incorporation of the system into the pythonNet code as modules consistent with the python paradigm:

```
mport system

class Box:
    def __init__(self):
        self.Length = 16
        self.Width = 4
```

```
    def __del__(self):
        if not self.IsDispose:
            self.Dispose()
```

Built modules are used in Accessing Namespace - Assemblies is replaced with import

Object methods, however, are often invoked within custom methods and EventHandlers. First a consideration of invocation within custom methods. The coding of custom method is every bit the same as coding regular python methods. So a methods named, say, CustomMethod coding begins with the general form as

```
def MethodName(self):
    return True
```

However, between the two bounding lines are inserted the rest of the codes guiding the behaviour of the objects, for instance an initialization of an object property can be codes as

```
def CustomMethod(self):
    frmObj.BeginUpdate()
    frmobj.property = initdata
    frmObj.EndUpdate
    return True
```

in which the common object of Windows Forms methods, BeginUpdate and EndUpdaten have both been invoked, for the purposes of ensuring the content update of the Form object.

Events and EventHandlers are coded somewhat differently,being coded with two additional standard arguments like so:

```
def EventHandlerName(self, sender, eventargs);
    return True
```

but even then, an EventHandler like the CustomMethod example given above can still be coded as that object methods are invoked as follows:

```
def EventHandlerName(self, sender, eventargs);
    if sender == frmObj:
        frmObj.BeginUpdate()
        frmobj.property = initdata
        frmObj.EndUpdate
    return True
```

Another significant difference is the if conditional statement that checks to be sure the code is being executed for the right sender object.Otherwise, methods and EventHandlers are coded as per the discretion of the developer to accomplish certain objectives

Within the context of the C# code translating doctrine then the basic PythnNet GUI code by which the Figs. 1.1 to 1.4 were produced is given here with understanding by the developer.

```
Using System import String
Using System.Collections import *
Using System.Globalization import *
Using Microsoft.Win32
Using win32file, win32pipe

Using System.Drawing import Color, Point, SizeF, Size
Using System.Windows.Forms import (Application, BorderStyle,
    Form,FormBorderStyle, Screen)
Using System.Windows.Forms import (ToolStripPanel,
    ToolStripContainer, ToolStrip, MenuStrip,
    ToolStripMenuItem, StatusBar, StatusBarPanel)
Using Microsoft.Win32
Using System.Collections import Generic
Using System.IO
```

```
Using System.Windows.Forms as WinsForms
Using System.Drawing as WinsDraw
Using System.Diagnostics as sysdiag
Using System.ComponentModel as syscompmodel
Using System.IO as sysio
Using Microsoft.Win32 as msw32
```

Within the context of the C# code translating doctrine then the basic Pythn-Net GUI code by which the Figs. 1.1 to 1.4 were produced is given here with understanding by the developer.

```
import clr
from System import String
from System.Collections import *
from System.Globalization import *
import Microsoft.Win32
import win32file, win32pipe

clr.AddReference('System.Data')
clr.AddReference('System.Drawing')
clr.AddReference('System.Windows')
clr.AddReference('System.Collections')
clr.AddReference('System.ComponentModel')
clr.AddReference('System.Globalization')
clr.AddReference('System.Windows.Forms')

from System.Drawing import Color, Point, SizeF, Size
from System.Windows.Forms import (Application, BorderStyle,
    Form,FormBorderStyle, Screen)
from System.Windows.Forms import (ToolStripPanel,
    ToolStripContainer, ToolStrip, MenuStrip,
```

```
    ToolStripMenuItem, StatusBar, StatusBarPanel)
from Microsoft.Win32 import *
from System.Collections import Generic
from System.IO import *
import System.Windows.Forms as WinsForms
import System.Drawing as WinsDraw
import System.Diagnostics as sysdiag
import System.ComponentModel as syscompmodel
import System.IO as sysio
import Microsoft.Win32 as msw32
```

4.7 Some Common C# Properties and Methods

As observed previously in Section 4.4.1, certain properties of the C# .Net objects are common to virtually all the Form objects and so maintaining templates coding of those properties and possible even methods and events would be helpful in aiding a developer to use preset values for immediate use even if subject to minor adjustments hence for the developer to plan to always code those first by default and the look for object specific codes to further specialize the object

The properties which have been collected together in Section 4.4.1 as common are coded presently for template use

```
guiobj = WinsForms.ClassObject()
guiobj.Name = aname
guiobj.Location =
guiobj.Width = 210
guiobj.Height = 25
guiobj.Point = WinsForms.Point(x, y)
guiobj.Size = WinsForms.Size(w,h)
guiobj.Alignment =
```

Obviously the assignment to these properties could be numerical of by the use of built in functions where available. These properties then are ones that should always get set in the design of the Windows Forms Objects.

Chapter 5

Initialization Logical Designs

5.1 PythonNet Logical Designs

The Logical Design[7] of pythonNet GUI may be grouped as of two scales: Static Logical design and Dynamic Logical Design. the former is generally implemented as the Initialization of the properties of the Forms objects, while the latter is often of the category of EventHandlers being the response to in-use dynamics. Accordingly, the Initialization codings focus generally on the Common Properties, including assignments to such objects as ComboBox, and also the execution of certain methods that effect initializations, etc

Developing .Net GUI objects therefore focuses on the development of the initialization coding. In particular the common properties are implemented for each of the objects first and then further specialized with the logical design of the object-specific properties. The coding of Form objects Logical design therefore starts off with the Initialization programming and is the object of the presentation in this chapter. In particular the coding of the most frequently used the GUI objects initializations are presented with insightfulness.

The pythonNet Logical Designs generally is always performed on a Windows Forms object as derived. The bare minimum Windows forms on which the various form objects can nbe developed is represented by the below code which invariably includes the C# "Using" translations as given Section ??:

```
#//Global call interface function
from __future__ import division
import sys #for exit
import socket #for sockets
import math
import time
#import pylab
#import matplotlib.pyplot as plt
from numpy import *
#from scipy.integrate import simps
import sympy
import ast
import threading
import pywintypes
import win32ui
import win32con #constants
#import pywin32
import ctypes
#//these are application related class-modules
import os as os
import fnmatch, re
from pathlib import Path
#translations of C\# components into pythonnet

import clr
from System import String
from System.Collections import *
from System.Globalization import *
import Microsoft.Win32
```

```
import win32file, win32pipe

clr.AddReference('System.Data')
clr.AddReference('System.Drawing')
clr.AddReference('System.Windows')
clr.AddReference('System.Collections')
clr.AddReference('System.ComponentModel')
clr.AddReference('System.Globalization')
clr.AddReference('System.Windows.Forms')

from System.Drawing import Color, Point, SizeF, Size
from System.Windows.Forms import (Application, BorderStyle,
    Form, FormBorderStyle, Screen)
from System.Windows.Forms import (ToolStripPanel,
    ToolStripContainer, ToolStrip, MenuStrip, ToolStripMenuItem,
    StatusBar, StatusBarPanel)
from Microsoft.Win32 import *
from System.Collections import Generic
from System.IO import *
import System.Windows.Forms as WinsForms
import System.Drawing as WinsDraw
import System.Diagnostics as sysdiag
import System.ComponentModel as syscompmodel
import System.IO as sysio
import Microsoft.Win32 as msw32

class pyNetBasicProgram(Form):
    def __init__(self):
        Form.__init__(self)
        self.IsMdiContainer = True
        self.Text = "Customer Sourcer"
```

```
            self.FormBorderStyle = FormBorderStyle.FixedDialog
            screenSize = Screen.GetWorkingArea(self)
            self.Height = screenSize.Height
            self.Width = screenSize.Width
            self.panelHeight = self.ClientRectangle.Height / 2
            self.CenterToScreen()
            self.appdir = os.getcwd()
            self.LayoutMdi(3)
            return True

        def GUIFormShow(self):
            self.Show()
            return True

    def __del__(self):
            pass
```

Analyzing the above code, the arrangement can be summarized generally as follows:

1. Python modules import
2. Windows modules import
3. FormObject module file

So then it is reasonable for compactness and focus to repackage the basic code into three files groups.

First, in the grouping is the python modules files, being named guipymod.py, with the designated files being defined as:

```
#FileName: guipymod.py
#//Global call interface function
```

```
from __future__ import division
import sys #for exit
import socket #for sockets
import math
import time
from numpy import *
import sympy
import ast
import threading
import pywintypes
import win32ui
import win32con #constants
import ctypes
#these are application related class-modules
import os as os
import fnmatch, re
rom pathlib import Path
```

Second in the grouping is the python modules files, being named guiwinsforms.py, with the designated files being defined as:

```
import clr
from System import String
from System.Collections import *
from System.Globalization import *
import Microsoft.Win32
import win32file, win32pipe
#import System.Windows.Controls.AxAcroPDFLib
#import AxAcroPDF1

clr.AddReference('System.Data')
clr.AddReference('System.Drawing')
```

```
clr.AddReference('System.Windows')
clr.AddReference('System.Collections')
clr.AddReference('System.ComponentModel')
clr.AddReference('System.Globalization')
clr.AddReference('System.Windows.Forms')
#clr.AddReference('System.Windows.Controls.AxAcroPDFLib')
#clr.AddReference('Microsoft.Win32')

from System.Drawing import Color, Point, SizeF, Size
from System.Windows.Forms import (Application, BorderStyle,
    Form, FormBorderStyle, Screen)
from System.Windows.Forms import (ToolStripPanel,
    ToolStripContainer, ToolStrip, MenuStrip, ToolStripMenuItem,
    StatusBar, StatusBarPanel)
from Microsoft.Win32 import *
from System.Collections import Generic
from System.IO import *
import System.Windows.Forms as WinsForms
import System.Drawing as WinsDraw
import System.Diagnostics as sysdiag
import System.ComponentModel as syscompmodel
import System.IO as sysio
import Microsoft.Win32 as msw32
```

Finally the third group of the codes are set into a function and saved as guilayout.py and are as given by the file:

```
def SetUpGUIForm(self):
    Form.__init__(self)
    self.IsMdiContainer = True
    self.Text = "Customer Sourcer"
    self.FormBorderStyle = FormBorderStyle.FixedDialog
```

```
        screenSize = Screen.GetWorkingArea(self)
        self.Height = screenSize.Height
        self.Width = screenSize.Width
        self.panelHeight = self.ClientRectangle.Height / 2
        self.CenterToScreen()
        self.appdir = os.getcwd()
        self.LayoutMdi(3)
        return True
```

and save this in a folder named formbase, for reuse as template code for form design, of course, subject to further modifications as deemed necessary, but the basic GUI iInterface code now being reduced as such:

```
import guipymode
from guipymod import *
import huwinsforms
from guiwinsforms import *
import formbase.guilayout
from formbase.guilayout import SetUpGUIForm

class pyNetBasicProgram(Form):
    def __init__(self):
        self.SetUpGUIForm(self)
        self.GUIFormShow()

    def GUIFormShow(self):
        self.Show()
        return True

    def __del__(self):
        pass
```

So then even without further additions to the contents all the GUI objects development coding guidance will be embedded within the above code in more effective demonstration of the development.

5.2 Prefixing GUI Objects Monikers

A good practice of coding GUI objects is to prefix the objects with a reference tag to enable identification of the object type in addition to the name of the object, so at debugging it is easier to keep in the mind the type of object being sought for editing.

Although, a developer is free to choose as pleases the choice of prefixes to the GUI objects such as adopted in the presentations and as recommended are as listed. Indeed, the examples given presently on the coding of the objects

Prefix	Object-Type
lbl	Label
txt,txb	TextBox
btn	Button
cbb	ComboBox
lsb	ListBox
ltv, lsv	ListView
rtb	RichTextBox
trv	TreeView
prb	ProgressBar
grb	GroupBox
cnn, ctn	Container
tbc	TabbedDialog

are all prefixed for effects while using the moniker, Example, for every one of the objects.

5.3 Developing .Net Label

Label is a reserved space on the Form in which texts are usually written either for identification or for giving instructive guidance or explanation of a particular action to the user. as such Labels are very plentiful in the design of GUIs for applications and other uses.

As with all Windows objects the Label is also of a nested class of the Form class and as such it is an inner class of the form class and so create-specified by the standard inner classes instantiation method, as per given in Section 4.6, written as:

```
self.lblExample = WinsForms.Label()
```

The create-specifications upon fully coded for a label for display on a GUI, and implemented is as shown in Fig.5.1, referencing "Example Label" in some respect as instruction to user. Often these labels give guidance about inserting a textual material into an text-data entry space on the GUI although such entry space does not exist in this GUi example.

The codes for instantiation of an object of the Label class on the form , based on the common properties as listed in Section 4.7, in full contents are:

```
import guipymode
from guipymod import *
import huwinsforms
from guiwinsforms import *
import formbase.guilayout
from formbase.guilayout import SetUpGUIForm

class ExampleLabel(Form)
    def __init__(self):
        self.SetUpGUIForm(self)
        self.InstanceLabel()
        self.GUIFormShow()
```

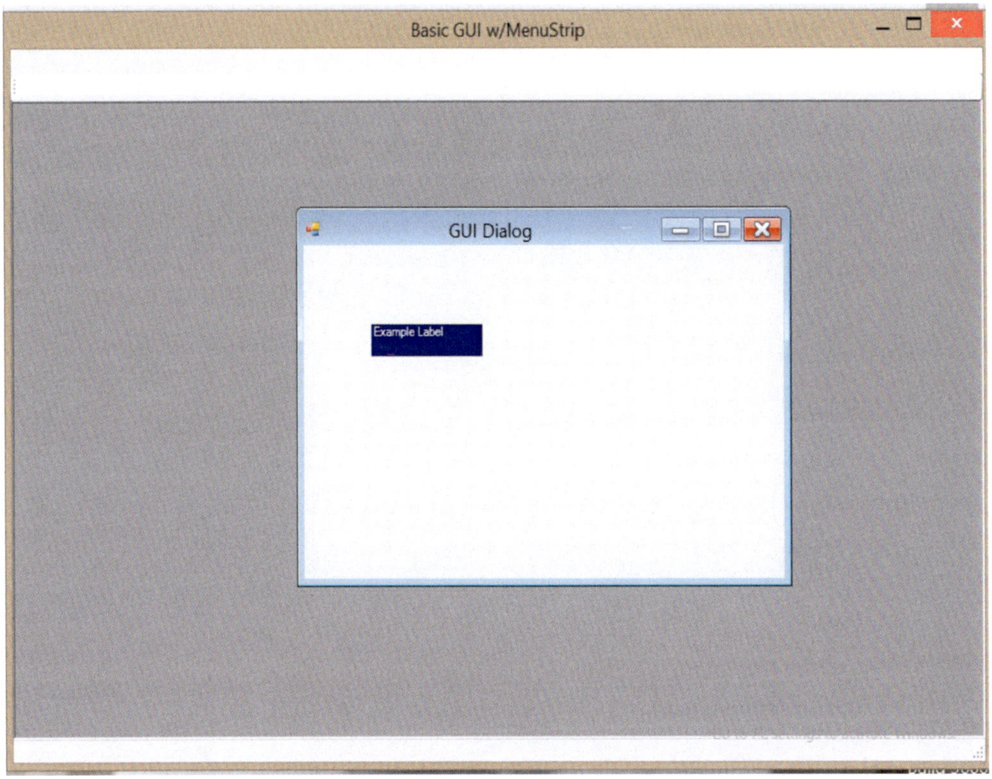

Figure 5.1: Label Example pytonNet GUI

```
def InstanceLabel(self):
    self.lblExample = WinsForms.Label()
    self.lblExample.Name = "lblExample"
    self.lblExample.Location = WinsDraw.Point(70, 75)
    self.lblExample.Height = 20
    self.lblExample.Width = 115
    self.lblExample.BackColor = Color.LightGray
    self.lblExample.ForeColor = Color.Black

    self.lblExample.Text = "Example Label"
```

```
            self.Controls.Add(self.lblExample)
            return True

    def GUIFormShow(self):
        self.Show()
        return True

    def __del__(self):
        pass
```

and with the implementation of the Text property code-line reflecting object type specific properties relative to the common properties. The Text property code is the mechanism by which the instructions are given with the Label object. The text for display by the Label object is usually given as assignment to the property at design time initialization:

```
    self.lblExample.Text = "Example Label"
```

However, the textual information or instructive guidance may also very well have been left as empty string, Str(""), and be inserted in the customary object initialization function with the inclusion of the codes;

```
    def InitializeLabel(self):
        self.lblExample.BeginUpdate
        self.lblExample.Text = str("Example Label")
        self.lblExample.EndUpdate
        return True
```

but then also including for such cases, the function invocations in design specifications of

```
self.InitializeLabel()
```

Alternatively, most such codes could also have been put in Custom and Helper functions, and in EventHandler coding; some examples are as follows

- Sample Custom Function code demonstrating use of the method: BeginUpdate() and EndUpdate();

```
def TextChangeMethod(self):
    self.lblExample.beginUpdate()
    self.lblExample.Text = ""
    self.lblExample.Text = str("New Text")
    self.lblExample.EndUpdate()
    return True
```

- Sample Click EventHandler code a similar change can also be induced by raising event as clickupon click of the label, and that could be written as:

```
def ClickEventhandler(self, sender, eventargs):
    if sender == self.lblExample:
        self.lblExample.beginUpdate()
        self.lblExample.Text = ""
        self.lblExample.Text = str("New Text")
        self.lblExample.EndUpdate()
    return True
```

Of course, of what implementation a developer adopts will depend on the circumstances prevailing and of the preference.

5.4 Developing .Net Button

Button is a reserved space on the Form that function, as it were an digital actuator akin to mechanical switches, and with texts usually written on them specifying the object of the actuation effectively giving instructive guidance or explanation of a particular action to the user. As such, Buttons are very plentiful in the design of GUIs including pythonNet GUIs.

As Windows Form object, the Button is also of a nested class of the Form class and as such it is an inner class of the Form class and so the standard inner classes instantiation method, as per given in Section 4.6 starts off the create-specification upon invocation, written as:

```
self.btnExample = WinsForms.Button()
```

The result of a fully specified Button create-code, for implementing a Button object in GUI is as shown in Fig.5.2, which is used to give an instruction to user

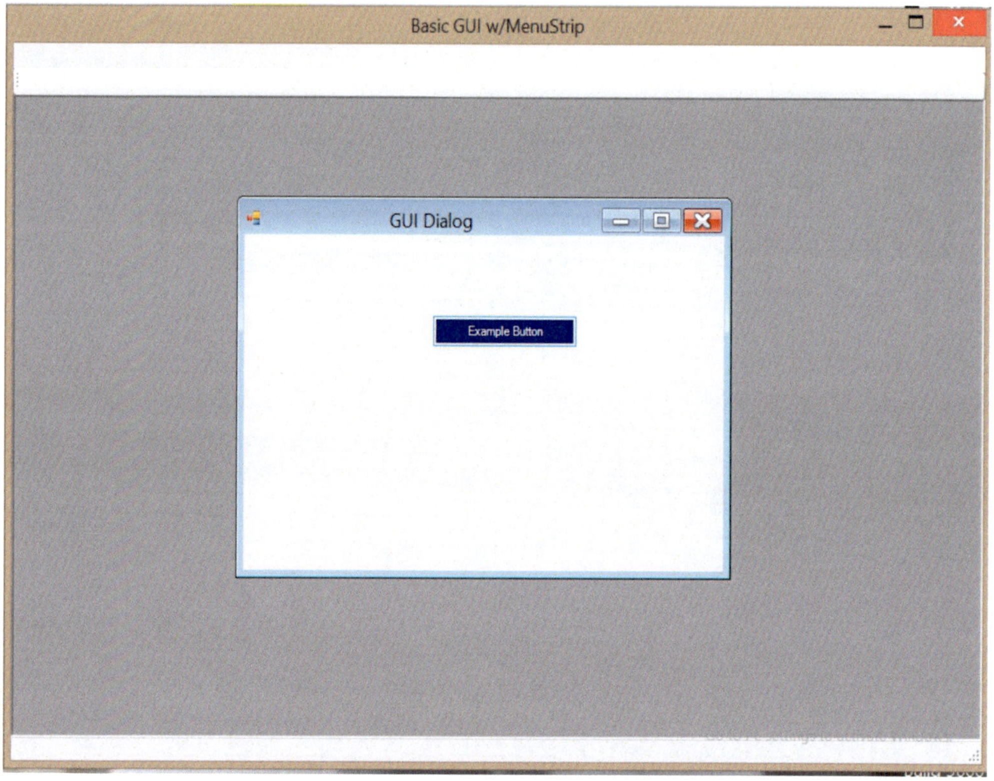

Figure 5.2: Button Example pythonNet GUI

about actuating the occurence of specific task as desired on the GUI.

The code for instantiation of an object of the Button class on the form is given in the full with respect to the common design properties of the form objects as listed in Section 4.7 is given by the codes-set:

```
import guipymode
```

```
from guipymod import *
import huwinsforms
from guiwinsforms import *
import formbase.guilayout
from formbase.guilayout import SetUpGUIForm

class ExampleButton(Form)
    def __init__(self):
        self.SetUpGUIForm(self)
        self.InstanceLabel()
        self.GUIFormShow()

    def InstanceButton(self):
        self.btnExample = WinsForms.Button()
        self.btnExample.Name = "btnExample"
        self.btnExample.BackColor = Color.LightGray
        self.btnExample.ForeColor = Color.Black
        self.btnExample.Height = 20
        self.btnExample.Width = 150
        self.btnExample.Location = WinsDraw.Point(195, 75)
        self.Controls.Add(self.btnExample)

        self.btnExample.Text = str("Send ID")
        self.btnExample.Click += self.ButtonClickHandler
        return True

    def ButtonClickHandler(self, sender, eventargs):
        #----direct some action here----
        return True

    def __del__(self):
```

```
        pass
```

including the two object specific property relative to the Common Properties which have assignments such as

```
self.tbnExample.Text = str("Button-Title")
self.btnExample.Click = EventHandlerName
```

The object specific property for the user-action Click-Event which defines the primary functionality of the object and its assignment:

```
self.btnExample.Click += self.ButtonClickHandler
```

imperative to the Button object that its is optional to a Label object. The EventHandler of the code which is also shown, generally embody codes directing that specific action be taken

```
def ButtonClickHandler(self, sender, eventargs):
        #----direct some action here----
        return True
```

The Text property of the Button is usually assigned to, at Design but textual information could also very well be changed dynamically in the EventHandler after the click, and preferring rigour also force the update of the title guidance as reflected:

```
def PerformAction(self, sender, eventargs):
    self.btnExample.BeginUpdate
    self.btnExample.Text = str("Action Performed")
    self.btnExample.EndUpdate
    return True
```

This action may be implemented in an effort to alert the end-user of the successful performing of the desired action.

However, the object specific methods of the Button object may also be coded inside Custom and Helper functions as well as EventHandler coding such as in the example functions:

- Sample Custom Function code demonstrating use of the method: BeginUpdate() and EndUpdate();

```
def TextChangeMethod(self):
    self.btnExample.BeginUpdate()
    self.btnExample.Text = ""
    self.btnExample.Text = str("Action Performed")
    self.btnExample.EndUpdate()
    return True
```

- Sample Click EventHandler code a similar change can also be induced by raising event as upon click of the Button, and that could be written as:

```
def DesignateEventHandler(self, sender, eventargs):
    if sender == self.btnDButton:
        self.TextChangeMethod()
    return True
```

The EventHandler name is completely arbitrary except to emhasize that the function is the choice function, and so the invocation could just have been of any other custom Method of choice of the developer

5.5 Developing .Net TextBox

TextBox is a reserved space on the Form in which texts can be added from externally by a user either for storage or insert into another GUI object; however, it can also be used as a label object if the property ReadOnly is set to True. In view of this multiple uses TextBoxes are very plentiful in the design of GUIs for applications and other uses.

As with all Windows Forms objects the TextBox is also of a nested class of the Form class and as such it is an inner class of the form class and so created by the standard inner classes instantiation method, as per given in Section 4.6 and is written as:

```
self.txtExample = WinsForms.TextBox()
```

The result of a fully coded a TextBox for display on a GUI is as shown in Fig.5.3, which is used to give an instruction to user about inserting a textual material into

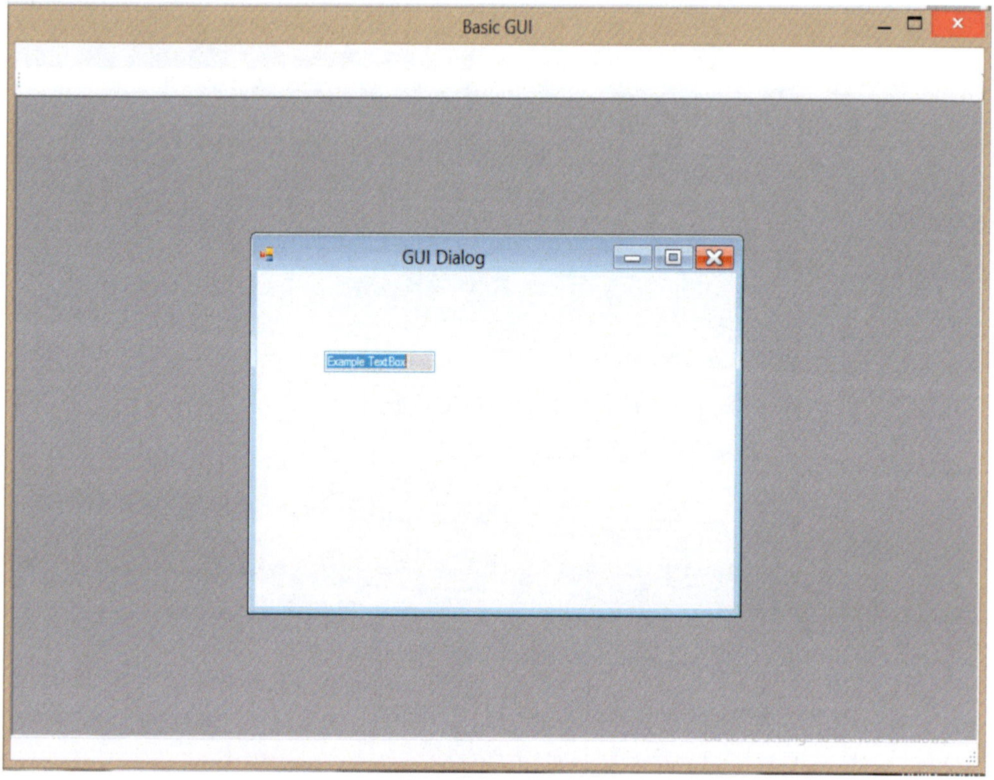

Figure 5.3: TextBox Example pythonNet GUI

an text entry space on the GUI although the object itself does not show up in this.

The code for instantiation of an TextBox object on the form is given in the full with respect to the properties of the form objects as listed in Section 4.7.

```
import guipymode
from guipymod import *
import huwinsforms
```

```
from guiwinsforms import *
import formbase.guilayout
from formbase.guilayout import SetUpGUIForm

class ExampleTextBox(Form)
    def __init__(self):
        self.SetUpGUIForm(self)
        self.InstanceTextBox()
        self.GUIFormShow()

    def InstanceTextBox(self):
        self.lblExample = WinsForms.Label()
        self.lblExample.Text = "Investor"
        self.lblExample.Location = WinsDraw.Point(70, 75)
        self.lblExample.Height = 20
        self.lblExample.Width = 115
        self.lblExample.BackColor = Color.LightGray
        self.lblExample.ForeColor = Color.Black
        self.datapanel.Controls.Add(self.lblExample)

        self.txtExample = WinsForms.TextBox()
        self.txtExample.BackColor = Color.LightGray
        self.txtExample.ForeColor = Color.Black
        self.txtExample.Height = 20
        self.txtExample.Width = 150
        self.txtExample.Location = WinsDraw.Point(195, 75)

        self.txtExample.Text = str(" ")
        self.Controls.Add(self.txtExample)
        return True
```

```
        def GUIFormShow(self):
            'self.Show()
            return True

        def __del__(self):
            pass
```

with the object-specific codes for the TextBox object relative to the Common Properties being as assigned

```
    self.txtExample.Text = str(" ")
    self.txtExample.Click
    self.txtAppendText
```

Being of functionality to permit entry of information into the software, the text display is usually implemented during use by writing to the reserved space as bounded. Effectively, the assignment of str(" ") could also very well have been left off without any adverse result.

Although the user Click-Event is not often emphasized, codes however could written in response towards further readying the object for the insertion of the information into the space. In such cases, additional EventHandler method are also co-invoked as shown

```
        self.txtExample.Click += self.NewEntryEventHandler
        return True

    def NewEntryEventHandler(self, sender, eventargs):
        if self.txtExample.OnGetFocus:
            self.txtExample.Text = str("")
        return True
```

which has the effect of clearing the object for the entry of the new information. Moreover, certain built-in methods could also be invoked to further customize the behavior The EventHandler of the object specific methods. Such methods are used inside Custom and helper functions and EventHandlers coding and some examples are as follows

- Sample Custom Function code demonstrating use of the method: BeginUpdate() and EndUpdate() together with certain KeyEvents[6]:

  ```
  self.txtDExample.KeyPress
  ```

  the KeyCode parameter of the KeyEventArgs can then be used in conjunction with other object state properties including.

- Sample Click EventHandler code related KeyEvents can also be induced by raising event as OnGotFocus and OnLostFous of the TextBox just before texts are keyed into it and just as the focus goes to another object,

  ```
  self.txtExample.OnGotFocus
  self.txtExample.OnLostFocus
  ```

  to develop more robust functionality of the TextBox; however, such are beyond the scope of a beginners

The property of AppendText is usually assigned to in Custom Methods or helper Function. In custom Method the loading of textual materials into the object is coded as follows:

```
self.txtExample.AppendText(uploadtext)
```

such that each upload is properly pre-formatted to be synchronized with the append operations, for better reading. Such application often obtains in the cases for which the TextBox object is also sized with the WinsForms.Size method, and wit the Width and Height property.

5.6 Developing .Net RichTextBox

RichTextBox is a reserved space on the Form in which large texts materials are usually written either for any purposes but allows text editing and advanced formatting such that even Richfield Texts could be loaded.

As with all Windows objects the RichTextBox is also of a nested class of the Form class and as such it is an inner class of the form class and so create-specification invoke consistent with the standard inner classes instantiation method, as per given in Section 4.6 starts off its implementation as written:

```
self.rtbExample = WinsForms.RichTextBox()
```

The result of a fully coded a RichTextBox for display on a GUI is as shown in Fig.5.4, which is used for extended textual materials for any purposes as may be necessary to user, such that large blocks of codes are entered into the software.

The code for instantiation an object of the RichTextBox class on the form is given in the full with respect to the properties of the form objects as listed in Section 4.7.

```
import guipymode
from guipymod import *
import huwinsforms
from guiwinsforms import *
import formbase.guilayout
from formbase.guilayout import SetUpGUIForm

class DRichTextBox(Form)
    def __init__(self):
        self.SetUpGUIForm(self)
        self.InstanceLabel()
        self.GUIFormShow()

    def InstanceLabel(self):
```

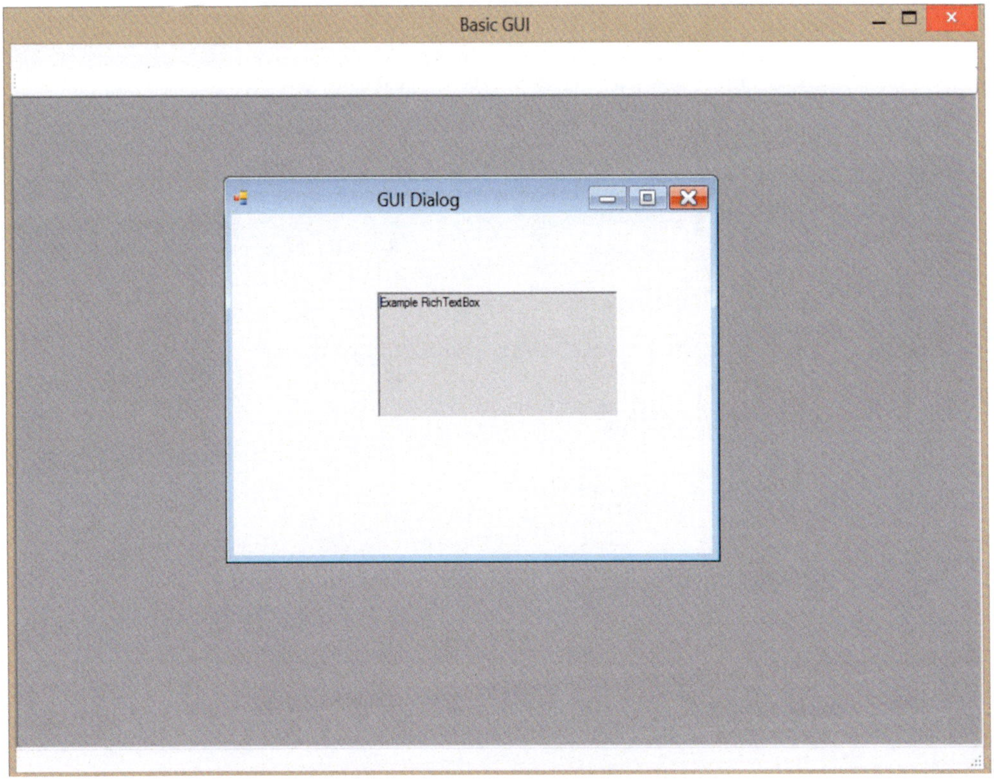

Figure 5.4: RichTextBox Example pythonNet GUI

```
        self.lblExample = WinsForms.Label()
        self.lblExample.Text = "Investor"
        self.lblExample.Location = WinsDraw.Point(70, 75)
        self.lblExample.Height = 20
        self.lblExample.Width - 115
        self.lblExample.BackColor = Color.LightGray
        self.lblExample.ForeColor = Color.Black
        self.Controls.Add(self.lblExample)

    def InstanceRichTextBox(self):
        self.rtbExample = WinsForms.RichTextBox()
```

```
            self.rtbExample.BackColor = Color.LightGray
            self.rtbExample.ForeColor = Color.Black
            self.rtbExample.Height = 120
            self.rtbExample.Width = 150
            self.rtbExample.Location = WinsDraw.Point(150, 75)

            self.rtbExample
            self.Controls.Add(self.rtbExample)
            return True

        def GUIFormShow(self):
            self.Show()
            return True

        def __del__(self):
            pass
```

Two object specific properties however are not included as they are built into the objects and so not usually explicitly coded by the developer. These two object specific properties:

```
self.rtbExample.Text
self.rtbExample.Click
```

yet are occasionally coded by developer when loading large textual materials into the software. The text material being required to be continually expanded with additional materials are often coded in a Helper Functions as with:

```
def WriteText(self, txtmat):
    self.rtbExample.BeginUpdate
    self.rtbExample.Text += txtmat
    self.rtbExample.EndUpdate
    return True
```

which uses the string concatenation operations of "+=" to enable the upload of the materials.

However this could also very well be changed dynamically after the click in the EventHandler and preferring rigour also force the update functions. As noted most such methods are used inside Custom and Helper functions and Eventhandler coding and some examples are as follows

- Sample Custom Function code demonstrating use of the method: BeginUpdate() and EndUpdate();

```
def TextChangeMethod(self):
    self.rtbExample.BeginUpdate()
    self.rtbExample.Text = ""
    self.rtbExample.Text = str("New Text")
    self.rtbExample.EndUpdate()
    return True
```

- Sample Click EventHandler code a similar change can also be induced by raising event as clickupon click of the RichTextBox, and that could be written as:

```
def ClickEventhandler(self, sender, eventargs):
    if sender == self.rtbExample:
        self.rtbExample.BeginUpdate()
        self.rtbExample.Text = ""
        self.rtbExample.Text = str("New Text")
        self.rtbExample.EndUpdate()
    return True
```

However, a third object-specific property of the RichTextBox object which is not built-in that must be explicitly coded by the developer enables certain editing functionalities that the object is required to provide, and is a ContextMenuStrip:

```
self.rtbExample.ContextMenuStrip = ContextMenuObject
```

of which the ContextMenuObject must be explicitly developed already and exist somewhere with the class object of the RichTextBox create-specifications, The specifics of this subject matter are addressed in Chapter 8, Section 8.7 with the full codes developed.

5.7 Developing .Net ProgressBar

ProgressBar is a reserved space on the Form in which the performing state of an ongoing task is being monitored in real-time, and thereby giving the user instructive guidance or explanation of a particular action. ProgressBar are very often used to inform on the state of information of an installation of software , downloading of code materials, or the uninstallation of software.

As with all Windows objects the ProgressBar class is also of a nested class of the Form class and as such, an inner class of the Form class hence its create-specification is the same as by the standard inner classes instantiation method, as per given in Section 4.6 and adapted here :

```
self.prbExample = WinsForms.ProgressBar()
```

The ProgressBar for display on a GUI as shown in Fig.5.5, is the instantiation coded fully consistent with the guidance regarding the evolution of an ongoing activity; and as such the user may effective infer the state of progression of the ongoing task on the GUI on which the developer is engaged, even although the task object does not show up in itself.

The code for instantiation an object of the ProgressBar class on the form is given in the full with respect to the properties of the form objects as listed in Section 4.7.

```
import guipymode
from guipymod import *
import huwinsforms
```

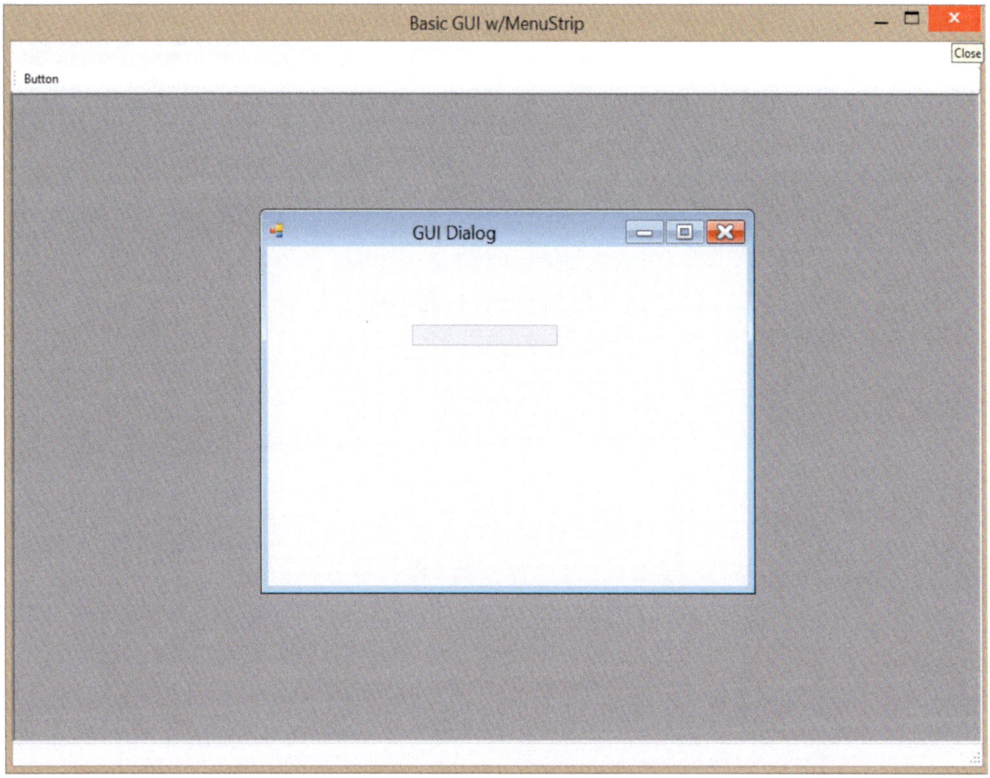

Figure 5.5: ProgressBar Example pythonNet GUI

```
from guiwinsforms import *
import formbase.guilayout
from formbase.guilayout import SetUpGUIForm

class ExampleProgressBar(Form)
    def __init__(self):
        self.SetUpGUIForm(self)
        self.InstanceDProgBar()
        self.GUIFormShow()

    def InstanceDProgBar(self):
```

```
        self.lblExample = WinsForms.Label()
        self.lblExample.Location = WinsDraw.Point(70, 75)
        self.lblExample.Height = 20
        self.lblExample.Width = 115
        self.lblExample.BackColor = Color.LightGray
        self.lblExample.ForeColor = Color.Black
        self.Controls.Add(self.lblExample)

        self.prbExample = WinsForms.ProgressBar()
        self.prbExample.BackColor = Color.LightGray
        self.prbExample.ForeColor = Color.Black
        self.prbExample.Height = 20
        self.prbExample.Width = 150
        self.prbExample.Location = WinsDraw.Point(195, 75)
        self.prbExample.Minimum = 1
        self.prbExample.Maximum = 100
        self.prbExample.Step = 1
        self.prbExample.Style = 1
        self.Controls.Add(self.prbExample)
        return True

    def GUIFormShow(self):
        self.Show()
        return True

    def __del__(self):
        pass
```

the object-specific codes for the ProgressBar obj are the following and can be determined from examples of the assignments to these properties. The object specific methods of the ProgressBar then are as follows:

```
self.prbExample.Minimum = 1
self.prbExample.Maximum = 100
self.prbExample.Step = 1
self.prbExample.Style = 2
```

As noted most such methods are used inside Custom and Helper functions and conditional change coding and some examples are as follows: Sample Custom Function code demonstrating use of the method: BeginUpdate() and EndUpdate();

```
def ProgressDisplayMethod(self):
    self.prbExample.BeginUpdate()
    self.prbExample.Maximum = 1000
    self.prbExample.Style = 2
    self.prbExample.EndUpdate()
    return True
```

which changes the Style and Maximum properties dynamically for effect that the developer may decide on. Even so such changes can be made rationally, say, to reflect when certain groups of software components are be executed or installed, and a code as could be used may be such as:

```
if self.pbr.Maximum == 25:
    self.ProgressBarChangeMethod()
return True
```

which, of course, implies that at the cumulative steps of 25, the components installation is due to change.

Of particular note is that the term ProgressBar is always aligned horizontally, however, there are two versions of this object actually: Horizontal ProgressBar and vertical ProgressBar, but horizontal ProgressBar is coding is rather developed in the above presentation.

5.8 Developing .Net ComboBox

ComboBox is a reserved space on the Form in which texts dataset is usually stored as options from which the user may make a choice from amongst the dataset. The dataset is usually made available in a DropDown listing.

As with all Windows objects the ComboBox is also of a nested class of the Form class and as such it is an inner class of the Form class and so its create-specification consistent with the standard inner classes instantiation method, as per given in Section 4.6, is written as:

```
self.cbbExample = WinsForms.ComboBox()
```

The complete code of design-specification of ComboBox for containing dataset on a GUI nis as shown in Fig.5.6, including essential object specific properties: to user about inserting a textual material into an text entry space on the GUI although the object itself does not show up in this.

The complete code of the design-specification for instantiation of an object of the ComboBox class on the form is given with full assignments to the Common Properties of the form objects as listed in Section 4.7 as:

```
import guipymode
from guipymod import *
import huwinsforms
from guiwinsforms import *
import formbase.guilayout
from formbase.guilayout import SetUpGUIForm

class DComboBox(Form)
    def __init__(self):
        self.SetUpGUIForm(self)
        self.InstanceLabel()
        self.GUIFormShow()
```

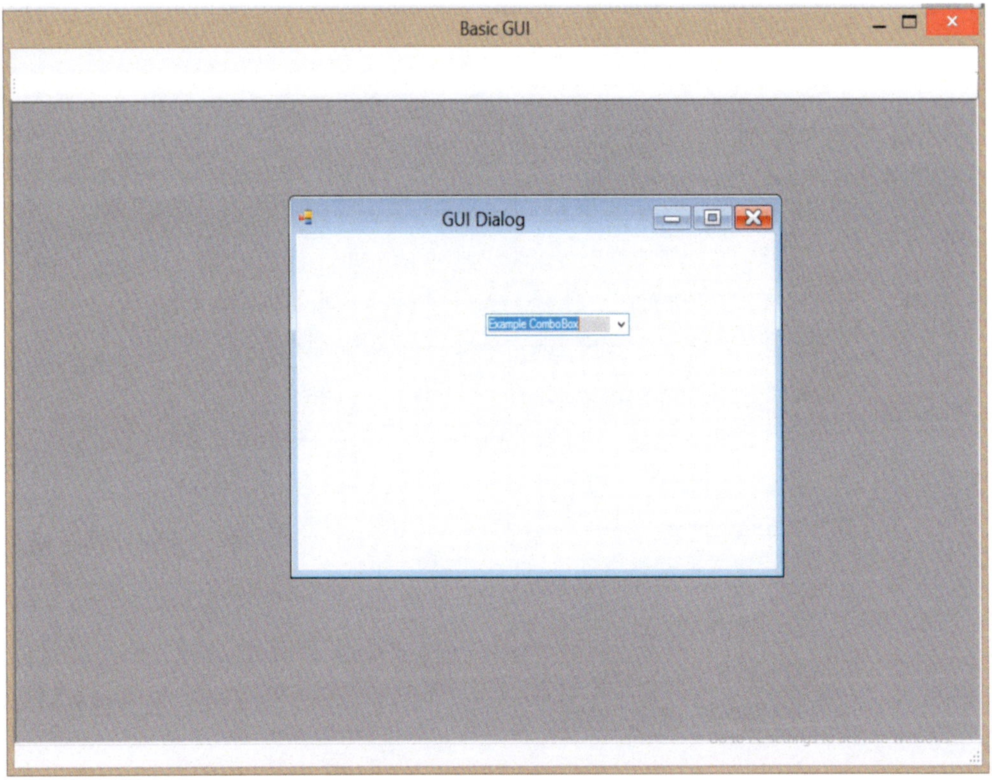

Figure 5.6: ComboBox Example pythonNet GUI

```
def InstanceComboBox(self):
    self.lblExample = WinsForms.Label()
    self.lblExample.Text = "Investor"
    self.lblExample.Location = WinsDraw.Point(70, 75)
    self.lblExample.Height = 20
    self.lblExample.Width = 115
    self.lblExample.BackColor = Color.LightGray
    self.lblExample.ForeColor = Color.Black
    self.Controls.Add(self.lblExample)

    self.cbbExample = WinsForms.ComboBox()
```

```
            self.cbbExample.BackColor = Color.LightGray
            self.cbbExample.ForeColor = Color.Black
            self.cbbExample.Height = 20
            self.cbbExample.Width = 150
            self.cbbExample.Location = WinsDraw.Point(195, 75)

            self.cbbExample.Items.Add("Make Choice")
            index = self.cbbExample.Items.FindString("Make Choice")
            self.cbbExample.SelectedIndex = Index
            self.cbbExample.Refresh()
            self.Controls.Add(self.cbbExample)
            return True

        def GUIFormShow(self):
            self.Show()
            return True

        def __del__(self):
            pass
```

with the object-specific properties being listed in the last batch of codes of the InstanceComboBox function:

```
    self.cbbExample.Items.Add("Make Choice")
    index = self.cbbExample.Items.FindString("Make Choice")
    self.cbbExample.SelectedItem = Index
    self.cbbExample.Refresh()
```

As an object for containing dataset of options, the insertion function for the text inserts into the ComboBox is usually conducted with the code:

```
    self.cbbExample.Items.Add("")
```

because in principle the dataset of strings, designated, say, as string(x) which the letter (x) runs through any number of numerals starting at 0, is effected as follows

```
self.cbbExample.BeginUpdate
self.cbbExample.Item.Add(string\textbf(x))
self.cbbExample.EndUpdate
```

The insert of the dataset could also have been made with explicit indexing of the dataset as examples of the assignments to these properties are as follows:

```
self.cbbExample.BeginUpdate
self.cbbExample.Item.Insert(index, string\textbf(x))
self.cbbExample.EndUpdate
```

where the Index specifies the preference of listing order of the content data of the objects.

However, such assignment property methods may also be coded within Custom and Helper functions, and EventHandler coding as these examples show:

- Sample Custom Function code demonstrating use of the method: BeginUpdate() and EndUpdate();

```
def DataSetChangeMethod(self):
    self.cbbExample.beginUpdate()
    self.cbbExample.Clear()
    self.cbbExample.Item.Add(string\textbf(x))
    self.cbbExample.EndUpdate()
    return True
```

- Sample Click EventHandler code a similar change can also be induced by raising event as clickupon click of the ComboBox, and that could be written as:

```
def ClickEventHandler(self, sender, eventargs):
    if sender == self.cbbExample:
        self.DataSetChangeMethod()
    return True
```

However the indexing order could also very well be preferred changed dynamically after the initialization, in which case index string-data pairs have to determined and re-sequenced. The efforts starts with the determination of the index associated with each string data for subsequent pairing, and the index is found as with the assignment:

```
index = self.cbbExample.Items.FindString("Make Choice")
```

Remarkably though the determination of the index may also be performed so that a specific data-string may be listed at the top as the default option by assignment of the index prioritize the position as

```
self.cbbExample.SelectedIndex = Index
```

An important object property not usually included in the design specification addresses the determination and retrieval of the option selected by a user

```
self.cbbExample.SelectedItem
```

and its use in collecting the selected option is implemented as follows

```
SelectedOption =  self.cbbExample.SelectedItem
```

after which the variable SelectedOption may be used by the developer as desired in other sections of the software.

5.9 Developing .Net ListBox

ListBox is a reserved space on the Form for use as data container in which text data are usually entered either as option choices or as packaging of related data for subsequent particular action of the user. The ListBox contained data are ordinarily visible to the end-user as listed items of string information.

As Windows object the ListBox is also of a nested class of the Form class and as such an inner class of the form class and therefore create-specified by the standard inner classes instantiation method, as per given in Section 4.6 as:

```
self.lsbExample = WinsForms.ListBox()
```

The GUI that obtains with fully coded ListBox object is as shown in Fig.5.7, displaying just one data entry although many ore could be added, at instantiation or by the user following display as shown; of course, the inserting of data by

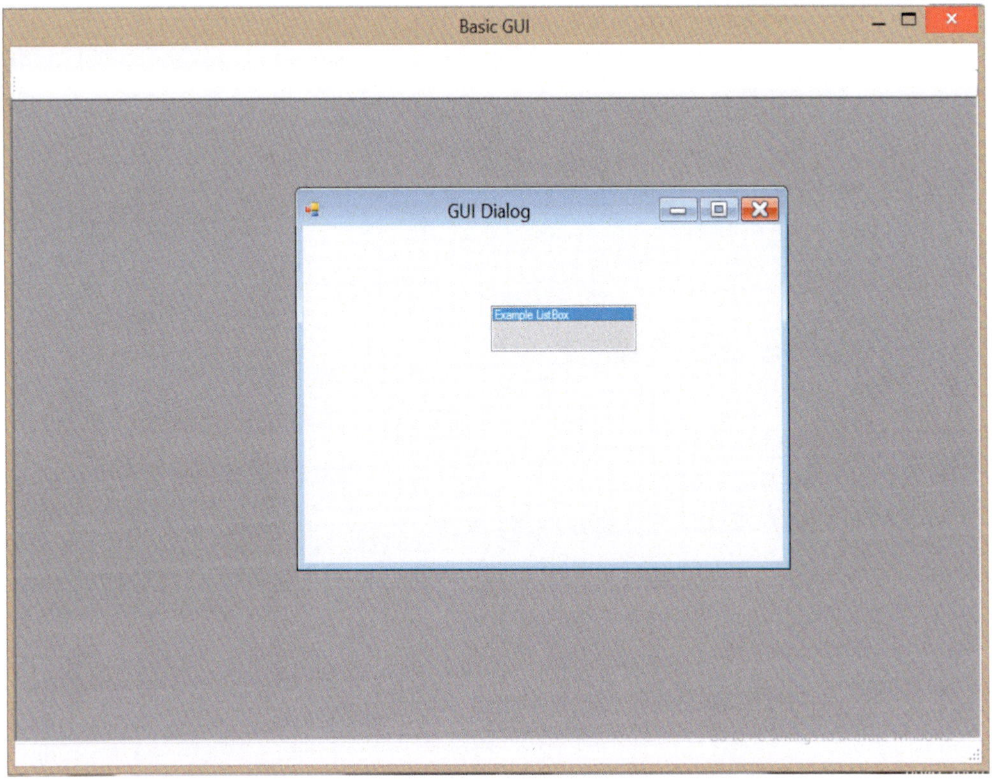

Figure 5.7: ListBox Example pythonNet GUI

the user requires a distinct TextBox Dialog onto which the textual materials are entered and then written to the ListBox.

Instantiating an object of the ListBox class on the form using the design specification of the common properties of the Form objects as listed in Section 4.7 is given by the full code:

```
import guipymode
```

```
from guipymod import *
import huwinsforms
from guiwinsforms import *
import formbase.guilayout
from formbase.guilayout import SetUpGUIForm

class DListBox(Form)
    def __init__(self):
        self.SetUpGUIForm(self)
        self.InstanceLabel()
        self.GUIFormShow()

    def InstaceDListBox(self):
        self.lblExample = WinsForms.Label()
        self.lblExample.Text = "Investor"
        self.lblExample.Location = WinsDraw.Point(70, 75)
        self.lblExample.Height = 20
        self.lblExample.Width = 115
        self.lblExample.BackColor = Color.LightGray
        self.lblExample.ForeColor = Color.Black
        self.Controls.Add(self.lblExample)

        self.lsbExample = WinsForms.ListBox()
        self.lsbExample.BackColor = Color.LightGray
        self.lsbExample.ForeColor = Color.Black
        self.lsbExample.Height = 20
        self.lsbExample.Width = 150
        self.lsbExample.Location = WinsDraw.Point(195, 75)

        self.lsbExample.Items.Add("Example ListBox")
        index = self.lsbExample.Items.FindString("Example ListBox")
```

```
            self.lsbExample.SelectedIndex= Index
            self.lsbExample.Refresh()
            self.Controls.Add(self.lsbExample)
            return True

        def GUIFormShow(self):
            self.Show()
            return True

        def __del__(self):
            pass
```

and has obvious similarity with the ComboBox object create specification code; the object-specific codes for the ListBox obj can be determined from examples of the assignments to these properties; and the object specific methods of the ListBox object are the last batch of codes of the InstanceListBox design codes:

```
self.lsbExample.Items.Add(data-item)
index = self.lsbExample.Items.FindString(toplist-data-item)
self.lsbExample.SelectedIndex= Index
self.lsbExample.Refresh()
```

As an object for containing dataset with some association, the method for the text inserts into the ListBox is usually conducted with the code:

```
self.lsbExample.Items.Add("")
```

However, as observed earlier, data could not be entered directly after the ListBox object has been instantiated and in display, and as such, the data will be added through a TextBox object of a GUI using the EventHandler of a Button object of the same GUI of the codes:

```
def EnterListBoxData(self, sender, eventargs):
    if sender == self.btnListBoxData:
```

```
            self.lsbExample.BeginUpdate
            textmat = self.txtListBoxData.Text
            self.lsbExample.Items.Add(textmat)
            self.lsbExample.EndUpdate
        return True
```

So then the Click Event of the Button object is raised and the EventHandler invoked, which collects the data entered in the TextBox and then assigns the data to the ListBox.

The insert of the dataset could also have been made with explicit indexing of the dataset as per the examples of the assignments to these properties as follows:

```
self.lsbExample.BeginUpdate
self.lsbExample.Item.Insert(index, pref-data)
self.lsbExample.EndUpdate
```

where the Index specifies the preference of listing order of the content data, pref-data, of the ListBox objects.

5.10 Developing .Net TreeView

TreeView is a reserved space on the Form for use as data container in which text data are usually entered by multiple Nested branches as either option choices or as packaging of related data for subsequent particular action of the user. The TreeView contained data are ordinarily visible or can be made readily visible to the end-user as listed in-tems of related Nested branches of string information.

As with all Windows objects the TreeView is also of a nested class of the Form class and as such it is an inner class of the Form class and as such its create-specification of the standard inner classes instantiation method, consistent with the guidance as per given in Section 4.6 is written as:

```
self.trvExample = WinsForms.TreeView()
```

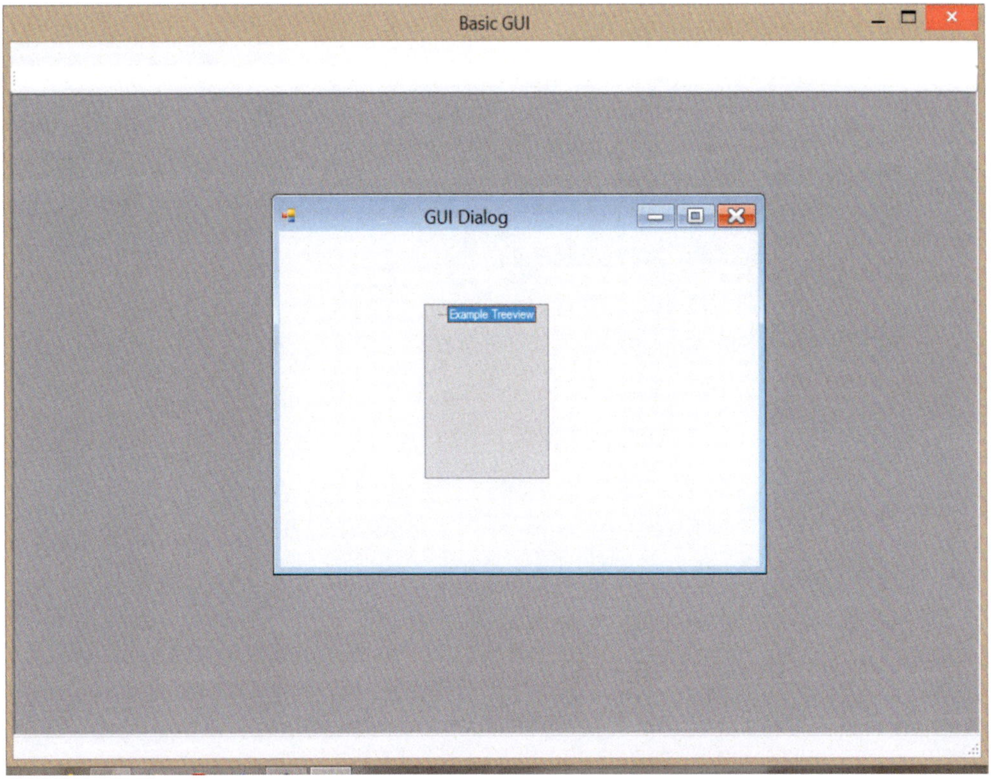

Figure 5.8: Treeview Example pythonNet GUI

The result of a fully coded a TreeView for display on a GUI is as shown in Fig.5.8, which is used to give an instruction to user about inserting a textual material into an text entry space on the GUI although the object itself does not show up in this.

The essential code of the design-specification for implementing an object of the Treeview class on the Form object with embedded Common Properties of the Form objects given with full assignement, as listed in Section 4.7 is as:

```
import guipymode
from guipymod import *
import huwinsforms
from guiwinsforms import *
```

```
import formbase.guilayout
from formbase.guilayout import SetUpGUIForm

class ExampleTreeview(Form)
    def __init__(self):
        self.SetUpGUIForm(self)
        self.InstanceLabel()
        self.GUIFormShow()

    def InstanceTreeView(self):
        self.lblExample = WinsForms.Label()
        self.lblExample.Text = "Investor"
        self.lblExample.Location = WinsDraw.Point(70, 75)
        self.lblExample.Height = 20
        self.lblExample.Width = 115
        self.lblExample.BackColor = Color.LightGray
        self.lblExample.ForeColor = Color.Black
        self.Controls.Add(self.lblExample)

        self.trvExample = WinsForms.TreeView()
        self.trvExample.Location = WinsDraw.Point(5, 71)
        self.trvExample.Width = self.Width/4 - 60
        self.trvExample.Height = self.Height - 200
        self.trvExample.HorizontalAlignment = WinsForms.
                                    HorizontalAlignment.Center
        self.trvExample.BackColor = Color.LightGray
        self.trvExample.ForeColor = Color.Black

        self.trvExample.ReadOnly = True
        self.trvExample.ScrollBars = WinsForms.ScrollBars.Vertical
        self.DesignTreeStructure()
```

```
        self.Controls.Add(self.trvExample)
        return True

    def GUIFormShow(self):
        self.Show()
        return True

    def __del__(self):
        pass
```

The data contents display of the TreeView is usually coded in a separate function, for purposes of compactness, and has the sample structure as given in the function:

```
def DesignTreeStructure(self):
    self.trvEmails.BeginUpdate()
    segscount = len(parfolders)
    segindexer = segscount
    nodeindex = -1
    for i in range(0, segscount):
        nodeindex += 1
        segsindex = segscount - segindexer
        parNode = WinsForms.TreeNode()
        parNode.Name = parfolders[i]
        parNode.Text = parfolders[i]
        #parNode.Index = i
        parNode.Tag = i
        self.trvExample.Nodes.Add(parNode)
        cfldercount = len(childfolders)
        #print(r"print salesreg[0] %4s"%(salesreg[0]))
        flderno = cfldercount
        #for j in range(0, terrcount):
        j = -1
```

```
            while flderno > 0:
                flderno -= 1
                territory = "Terr" + str("-") + str((cfldercount
                                                        - flderno))
                #print(r"print territory %34s"%(territory))
                childNode = WinsForms.TreeNode()
                childNode.Text = childfolders[j]
                childNode.Name = childfolders[j]
                nodeindex += 1
                j += 1
                childNode.Tag = nodeindex
                self.trvExample.Nodes[i].Nodes.Add(childNode)
        self.trvExample.EndUpdate()
        self.trvExample.Refresh()
        self.trvExample.CreateControl()
        return True
```

where two groups of data are used one as parent-data and the other as chlld-data. Each parent constitute a set of data-branch.

The object-specific codes for the TreeView object are mostly found in the Tree construction codes of DesignTreeStructure() and can be determined from examples of the assignments made in that function; the object specific methods of the Treeview object are as follows:

```
    parNode = WinsForms.TreeNode()
    parNode.Name = parfolders[i]
    parNode.Text = parfolders[i]
    parNode.Tag = i
```

which are the parent datasets that are used to create the parent node. then,of course, also these properties:

```
    childNode = WinsForms.TreeNode()
```

```
childNode.Text = childfolders[j]
childNode.Name = childfolders[j]
childNode.Tag = nodeindex
```

the use of these object-specific Properties and methods are obvious from the example code presented, however, ther are also other object specific properties and Methods that can be used for the same purposes. Some of these are:

```
self.trvExample.SelectedNode
self.trvExample.Nodes
```

the use of these object Properties is effective in the manipulation of data contained in a Treeview object. As a vehicle offering datasets as options of choice the capacity to select and then extract te chosen data is important. The task is met with the use of the SelectedNode property like so:

```
selnode = self.trvexample.SelectedNode
```

after which all the related data such as data, tag and parent of the SelectedNode can be got.. The property of the treeview object, Nodes, permits the construction of the nested data without exception, and allows a developer to code a recursive search through the nested data

5.11 Developing .Net ListView

ListView is a reserved space on the Form for use as data container in which text data sets are usually entered primarily for display in tabular structure for the perception of the user for such subsequent particular action of the user.

As with all Windows objects the label is also of a nested class of the Form class and as such it is an inner class of the form class and so created by the standard inner classes instantiation method, as per given in Section 4.6 and is written as:

```
self.ltvExample = WinsForms.ListView()
```

The result of a fully coded a label for display on a GUI is as shown in Fig.5.9, which is used to give an instruction to user about inserting a textual material into

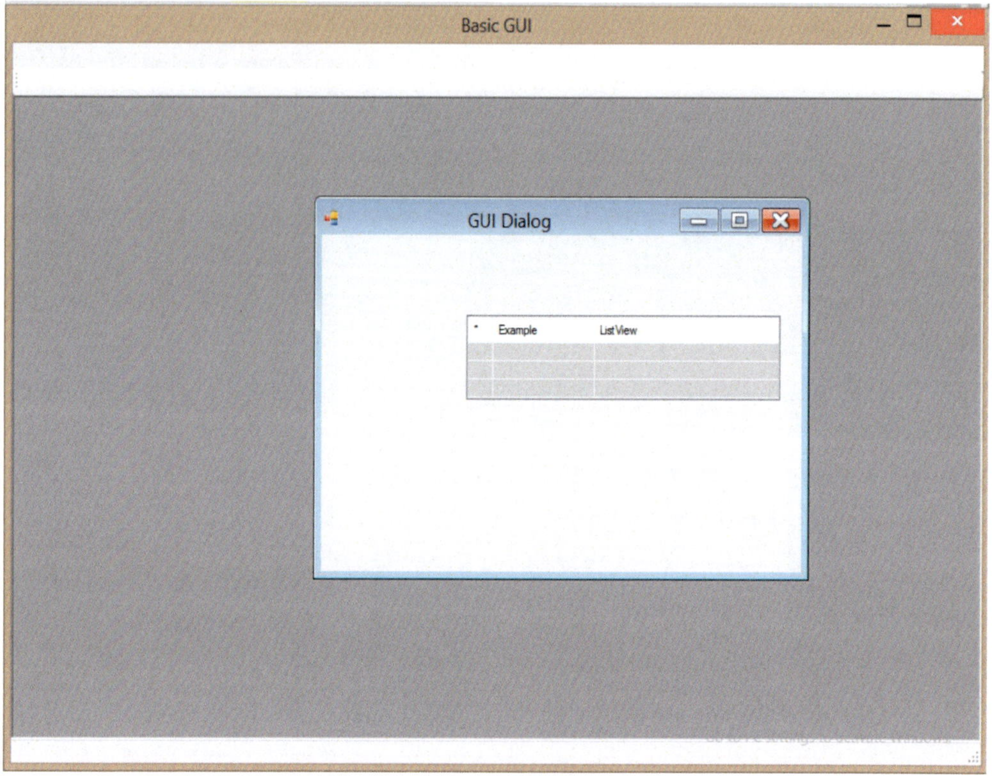

Figure 5.9: Listview Example pythonNet GUI

an text entry space on the GUI although the object itself does not show up in this.

The code for instantiation an object of the Listview class on the form is given in the full with respect to the properties of the form objects as listed in Section 4.7.

```
import guipymode
from guipymod import *
import huwinsforms
from guiwinsforms import *
```

```
import formbase.guilayout
from formbase.guilayout import SetUpGUIForm

class ExampleListView(Form)
    def __init__(self):
        self.SetUpGUIForm(self)
        self.InstanceLabel()
        self.GUIFormShow()

    def Instancelabel(self):
        self.lblExample = WinsForms.Label()
        self.lblExample.Text = "Investor"
        self.lblExample.Location = WinsDraw.Point(70, 75)
        self.lblExample.Height = 20
        self.lblExample.Width = 115
        self.lblExample.BackColor = Color.LightGray
        self.lblExample.ForeColor = Color.Black
        self.Controls.Add(self.lblExample)

        self.ltvExample = WinsForms.ListView()
        self.ltvExample.View = WinsForms.View.Details
        #self.ltvExample.CheckBoxes = True
        self.ltvExample.FullRowSelect = True
        self.ltvExample.GridLines = True
        self.ltvExample.BackColor = Color.LightGray
        self.ltvExample.ForeColor = Color.Black
        self.ltvExample.Size = WinsDraw.Size(int(self.Width * 3/4
                    - 60), int(self.Height/2 - 100))
        self.ltvExample.Location = WinsDraw.Point(5, 5)

        self.ltvExample.MultiColumn = True
```

```
        self.ltvExample.ColumnWidth = self.ltvExample.Width/2
        self.ltvExample.Columns.Add("*", 25, WinsForms.
                                    HorizontalAlignment.Left)
        self.ltvExample.Columns.Add("Sender", 105, WinsForms.
                                    HorizontalAlignment.Left)
        self.ltvExample.Columns.Add("Email Subject", 400,
                                WinsForms.HorizontalAlignment.Left)
        self.ltvExample.Columns.Add("Arrival-Date", 105,
                                WinsForms.HorizontalAlignment.Left)
        self.ltvDListViewItem = WinsForms.ListViewItem(varname,0)
        self.ltvProcSeqItem.SubItems.Add(str(seqindex))
        self.ltvProcSeq.Items.Add(self.ltvProcSeqItem)
        self.Controls.Add(self.ltvExample)
        return True

    def GUIFormShow(self):
        self.Show()
        return True

    def __del__(self):
        pass
```

The object-specific codes for the Listview object are determined from examples of the assignments to the design specification properties; and the object specific methods of the Listview object then are:

```
self.ltvExample.Columns.Add("*", 25, WinsForms.
                            HorizontalAlignment.Left)
self.ltvExample.Columns.Add("Sender", 105, WinsForms.
                            HorizontalAlignment.Left)
self.ltvExample.Columns.Add("Email Subject", 400, WinsForms.
                            HorizontalAlignment.Left)
```

```
self.ltvExample.Columns.Add("Arrival-Date", 105, WinsForms.
                              HorizontalAlignment.Left)
self.ltvDListViewItem = WinsForms.ListViewItem(varname,0)
self.ltvProcSeqItem.SubItems.Add(str(seqindex))
self.ltvProcSeq.Items.Add(self.ltvProcSeqItem)
```

The object specific properties and methods are of two groups because of he need to create a proper table set of data. The first group of properties and methods is the column group which specifies the column names, and consist of the following:

```
self.ltvExample.Columns.Add("*", 25, WinsForms.
                              HorizontalAlignment.Left)
self.ltvExample.Columns.Add("Sender", 105, WinsForms.
                              HorizontalAlignment.Left)
self.ltvExample.Columns.Add("Email Subject", 400, WinsForms.
                              HorizontalAlignment.Left)
self.ltvExample.Columns.Add("Arrival-Date", 105, WinsForms.
                              HorizontalAlignment.Left)
```

and although the Column property can be dynamically set at initialization, the preferred approach is to set the design-specification assignment at design.

The second group of properties and methods is the actual data group which specifies the data-sets associated with each column of the table that invariably obtains, and consist of the following:

```
self.ltvDListViewItem = WinsForms.ListViewItem(varname,0)
self.ltvProcSeqItem.SubItems.Add(str(seqindex))
self.ltvProcSeq.Items.Add(self.ltvProcSeqItem)
```

However, as observed earlier, data could not be entered directly after the Listview object has been instantiated and in display, and as such, the data will be added through a TextBox object of a GUI using the EventHandler of a Button object of the same GUI of the codes:

```
def EnterListViewData(self, sender, eventargs):
    if sender == self.btnListViewData:
        self.ltvExample.BeginUpdate
        textmat = self.txtListBoxData.Text\\
        self.ltvDListViewItem = WinsForms.ListViewItem(varname,0)
        delimiter = ","
        indx = 0
        while indx < len(txtmat):
            nIndx = textmat.find(delimiter, indx)
            datastr = textmat.slice[ indx, nIndx]
            self.ltvProcSeqItem.SubItems.Add(str(datastr))
            indx = nIndx
        self.ltvProcSeq.Items.Add(self.ltvProcSeqItem)
        self.ltvExample.EndUpdate
    return True
```

A comma separated set of data are entered into the TextBox object ,and the then the Click Event of the Button object is raised and the EventHandler invoked, which collects the data entered in the TextBox and then assigns the data to a variable, textmat. The variable is then tokenized and each token stored in datstr is then assigned to the column of the Listview object.

However, as a Data Container, there are other object-specific Properties and Methods that are used for the manipulation of the contained data, which are not immediately used in the create-design specification of the object, and as such not be in the class design. Those object-specific Properties and Methods include the following commonly used set:

```
self.ltvExample.SelectedItem
self.ltvExample.SelectedNode[0]
```

the data contained in a row is often accessed with the first Property and is used as follows:

```
rowdata =self.ltvExample.SelectedItem
colOneentry = rowdata[1]
col2entry = rowdata[2]
.
.
.
colNentry = rowdata[n]
```

The second object-specific Property is however used nly to collect the data in the very first column of the table of the choosen row, and is used as follows:

```
fcoldata = self.ltvExample.SelectedNode[0]
```

Although this presentation of the object-specific properties and Methods may not exhaustive most operations are performed with just these; and in fact because of the capacity to extract the entry per row, ven the re-arranging of row ca be effected readily.

Chapter 6

Container GUIs

6.1 GUI Layout Aids

There are, of course, also GUI objects which primary uses are the managing and organizing of the layout of the GUI objects of the previous Chapter 5. Some are quite simple and are not truly perceived by the viewer and others are quite noticeable by the end-user of software in they are implemented. Both types, however, are useful and serve their purposes, and as such deserve being developed when appropriate.

The development of those objects is addressed presently, starting with less obvious one to the most complex ones, and in order of their uses as need requires

6.2 Developing .Net Panel

Panel is an object of Windows Forms in which other GUI objects are placed in preferred arrangements . That they are container objects is reflected by the usual statement forcing the containing o the GUI objects being laid out relative to the geometry. The panel is also particularly useful in managing the background color of the GUI appearance.

As a Windows object Panel is also of a nested class of the Form class and as such it is an inner class of the form class and so created by the standard inner classes instantiation method, as per given in Section 4.6 and is written as:

```
self.uPanel = WinsForms.Panel()
```

The effect of a panel on the GUI background is as shown in Fig.6.1, as the section of the region that is purple being the color assigned to its background color

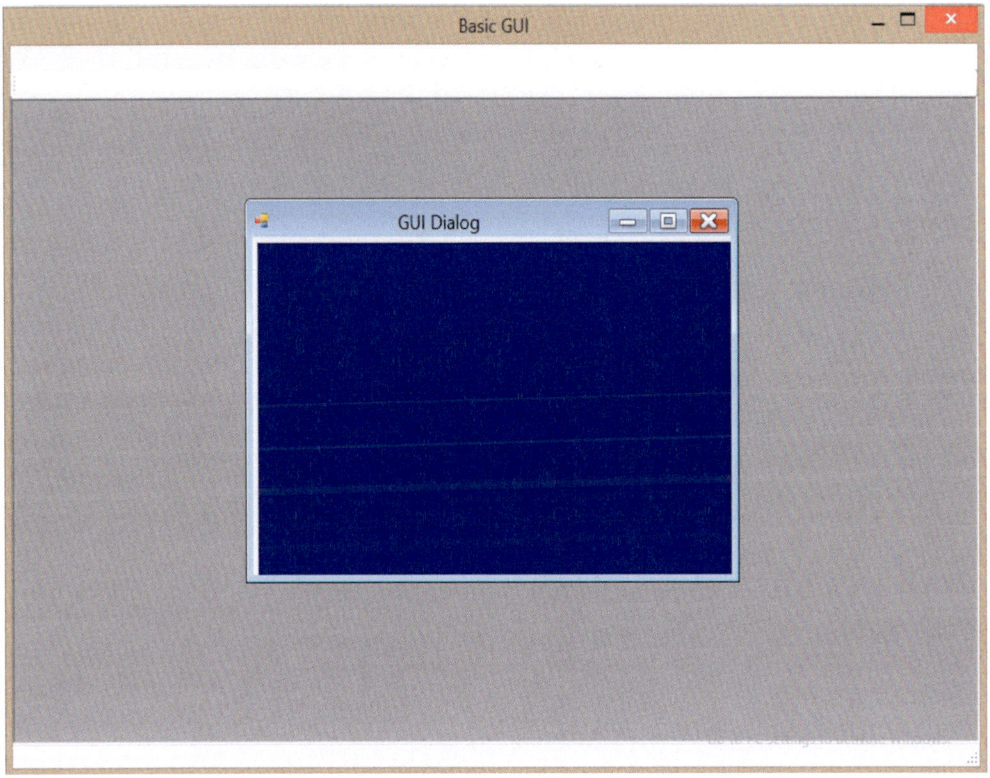

Figure 6.1: panel-container Example pyNetGUI

property.

The coding for an object of the panel class on the form is given in the full with respect to the properties of the form objects as listed in Section 4.7.

```
import guipymode
from guipymod import *
```

```
import huwinsforms
from guiwinsforms import *
import formbase.guilayout
from formbase.guilayout import SetUpGUIForm

class ExamplePanel(Form)
    def __init__(self):
        self.SetUpGUIForm(self)
        self.InstancePanel()
        self.GUIFormShow()

    def InstancePanel(self):
        self.uPanel = WinsForms.Panel()
        self.uPanel.Name = "Investor"
        self.uPanel.Location = WinsDraw.Point(0, 0)
        self.uPanel.Height = self.Height
        self.uPanel.Width = self.Width/2
        self.uPanel.BackColor = Color.Purple
        self.uPanel.ForeColor = Color.White
        self.uPanel.BorderStyle = BorderStyle.FixedSingle

        self.Controls.Add(self.uPanel)
        return True

    def GUIFormShow(self):
        self.Show()
        return True

    def __del__(self):
        pass
```

Of significant note is that the Panel will in fact not be displayed without the inclusion of the last line of code:

```
self.Controls.Add(self.uPanel)
```

before the return statement, because the parent-relationship of the Form object, in this case the *self* to the panel object , *uPanel* is defined by that association code.

The width of the panel has been sized as half the width of the GUI form object to elicit the background managing use of the panel; however, for real applications the width and height of the panel are set according to the preferences of the developer.

The color aspect of the panel can also e used to demarcate a Winds Forms object into section and with each section holding holding a different set of GUI objects. Indeed, the demarcating object can still be achieved without the background colour effect, although such makes the demarcation more stark or obvious.

6.3 Developing .Net GroupBoxes

GoupBox is a allocation of bounded space on the display region of the GUI Form object. The GroupBox is actually a second tier container in the sense that it can it is often itself contained in a Panel. The use of the GroupBox as such is mostly for containing similar GUI objects of multiplicity of count on the same GUI Display. ce or explanation of a particular action to the user. The GroupBox permits naming identification reflecting the objects contained within its bounded space for better informing of the end-user of the GUI of which it is also a component.

The GroupBox as with all Windows objects is also of a nested class of the Form class and as such is so created by the standard inner classes instantiation method, as per given in Section 4.6 and is written as:

```
self.grbExample = WinsForms.GroupBox()
```

The fully code-designed GroupBox as display on GUI is as shown in Fig.6.2, which informs the end- user about its containing GUI Objects. The code for

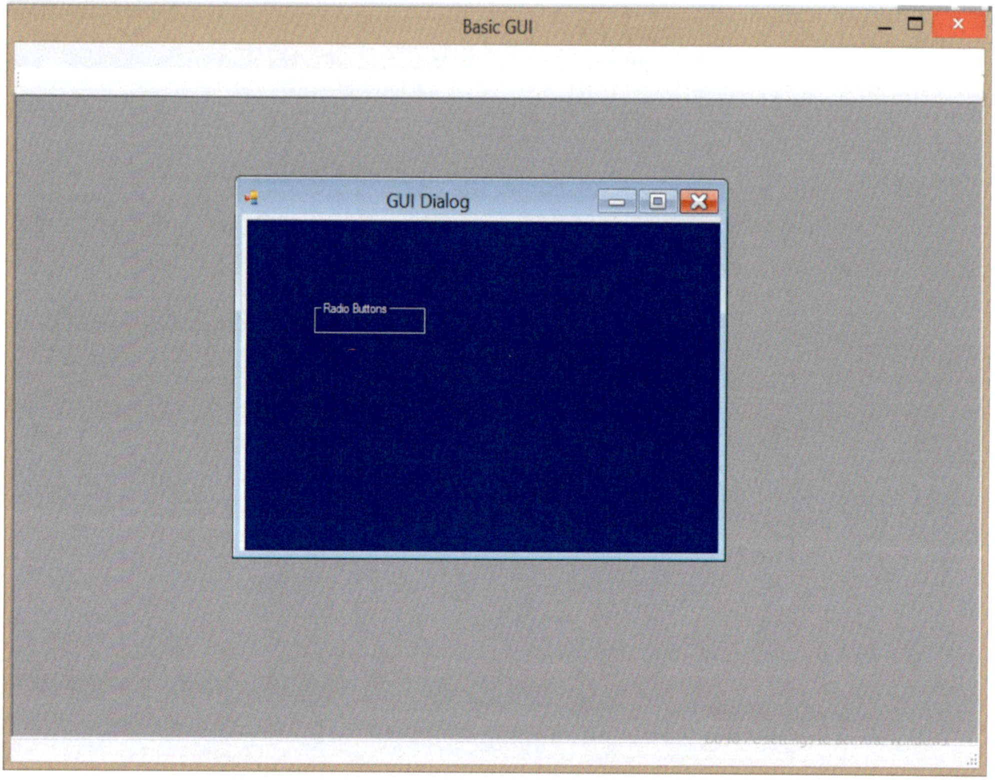

Figure 6.2: GroupBox Example pythonNet GUI

instantiating an object of the GroupBox class on the Form object is given in the full with respect to the properties of the Form objects as listed in Section 4.7.

```
import guipymode
from guipymod import *
import huwinsforms
from guiwinsforms import *
import formbase.guilayout
from formbase.guilayout import SetUpGUIForm

class ExampleGroupBox(Form)
    def __init__(self, parent):
```

```
        Form.__init__(self)
        self.MdiParent = parent
        self.parent = parent
        self.Text = "GUI Dialog"
        self.Height = self.parent.Height/2
        self.Width = self.parent.Width/2
        self.Location = WinsDraw.Point(int(self.parent.Width/4),
                                       int(self.parent.Height/4))

        self.InstancePanel()
        self.InstanceGroupBox()

        self.uPanel.Controls.Add(self.grbExample)
        self.Controls.Add(self.uPanel)

        self.GUIFormShow()

    def InstancePanel(self):
        self.uPanel = WinsForms.Panel()
        self.uPanel.Name = "ExamplePanel"
        self.uPanel.Location = WinsDraw.Point(5, 5)
        self.uPanel.Height = self.Height - 10
        self.uPanel.Width = self.Width - 10
        self.uPanel.BackColor = Color.Navy
        self.uPanel.ForeColor = Color.White
        self.uPanel.BorderStyle = BorderStyle.FixedSingle
        return True

    def InstanceGroupBox(self):
        self.grbExample = WinsForms.GroupBox()
        self.grbExample.Text = "Radio Buttons"
```

```
        self.grbExample.Location = WinsDraw.Point(70, 75)
        self.grbExample.Height = 30
        self.grbExample.Width = 115
        self.grbExample.BackColor = Color.Navy
        self.grbExample.ForeColor = Color.White
        self.grbExample.FlatStyle = WinsForms.FlatStyle.Standard
        return True

    def GUIFormShow(self):
        self.Show()
        return True

    def __del__(self):
        pass
```

Panel object is included here by as the default GUI object container. Significantly, the GroupBox object continues to be an abstract construction or a set of design specifications for implementing GroupBox contained in the panel which implementation is directed by the code

```
self.uPanel.Controls.Add(self.grbExample)
```

and it is the mounting of the specifications set on the panel object that makes the object accessible in memory.

6.4 Developing .Net Container

Container sections the Form object for the unique use for different sub-applications. Although, panels could be used to also section Forms objects, Container has the special advantage of sections the codes as well, enabling the compartmentalizing of the codes according to the contain in which the object is placed.

As with all Windows objects the Container is also of a nested class of the Form class and as such it is an inner class of the form class and so create-specified by the standard inner classes instantiation method, as per given in Section 4.6 and is written as:

```
self.cnnExample = WinsForms.SplitContainer()
```

which specifies a pair of objects of Windows Forms without actually yet dividing the Derived Form Object on which the pair Container-objects are yet to be prescribed. In any event, the fully codes-specified pair Container-objects when implemented on the Class derived form object display in the GUI as shown in Fig.6.3, which is best elicited with the mounting of one Panel object per Container-object

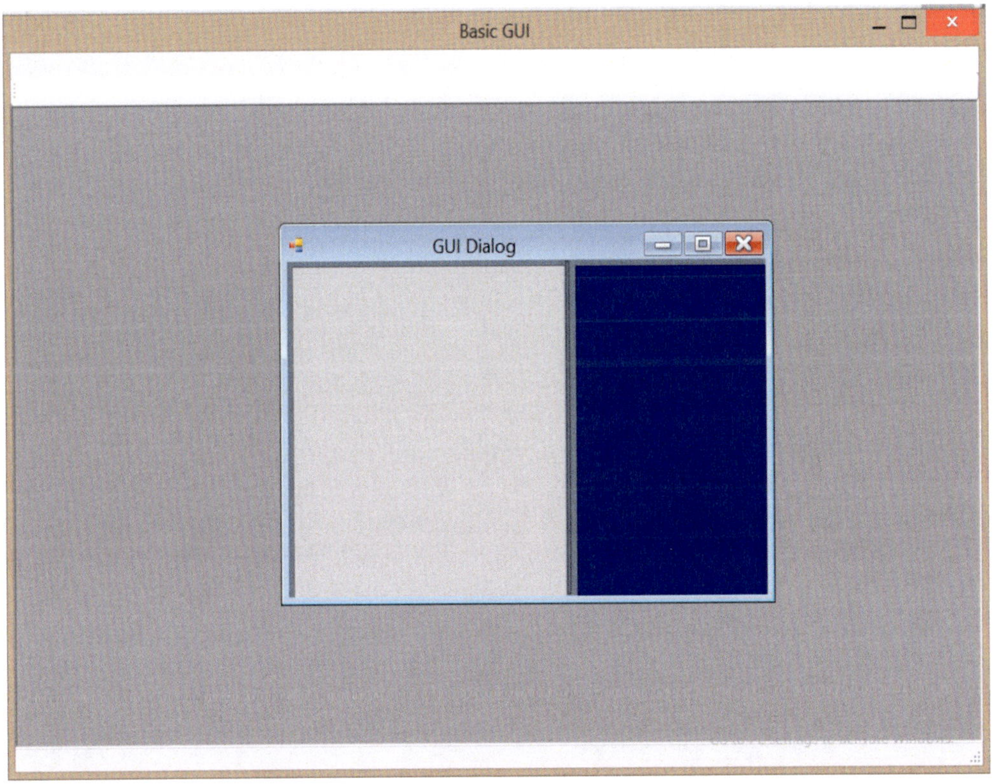

Figure 6.3: Container Example pyNet GUI

instructive of the user about the distinctness of two parts of the the same window view. The Container-Objects pair are reflected by the blue and LightGray colored parts of the Forms object

The code for specification and instantiation an object of the Container class on a Windows Form is with respect to the properties of the Form objects as listed in Section 4.7. as adapted given in the full

```
import guipymode
from guipymod import *
import huwinsforms
from guiwinsforms import *
import formbase.guilayout
from formbase.guilayout import SetUpGUIForm

class ExampleContainer(Form)
    def __init__(self, parent):
        Form.__init__(self)
        self.MdiParent = parent
        self.parent = parent
        self.Text = "GUI Dialog"
        self.Height = self.parent.Height/2
        self.Width = self.parent.Width/2
        self.Location = WinsDraw.Point(int(self.parent.Width/4),
                                       int(self.parent.Height/4))

        self.InstanceContainer()
        self.InstanceLeftPanel()
        self.InstanceRightPanel()

        self.readerContainer.Panel1.Controls.Add(self.lPanel)
        self.readerContainer.Panel2.Controls.Add(self.rPanel)
        self.Controls.Add(self.readerContainer)
        self.readerContainer.ResumeLayout(False)
        self.ResumeLayout(False)
```

```
        self.GUIFormShow()

    def InstanceContainer(self):
        self.Locked = False
        """ design specification of the Container objects..."""
        self.readerContainer = WinsForms.SplitContainer()
        self.readerContainer.SuspendLayout()
        self.SuspendLayout()

        self.readerContainer.Orientation = WinsForms.
                                    Orientation.Vertical
        self.readerContainer.SplitterDistance = 80
        self.readerContainer.SplitterIncrement = 5
        self.readerContainer.Location = WinsDraw.Point(0,0)
        self.readerContainer.BackColor = Color.LightSlateGray
        self.readerContainer.Size = WinsDraw.Size(
                int(self.Width/4 + 10), int(self.Height))
        self.readerContainer.SplitterWidth = 6
        self.readerContainer.Dock = WinsForms.DockStyle.Fill

        self.leftwin = self.readerContainer.Panel1
        self.readerContainer.Panel1.Name = "leftwin"
        #self.leftwin.Dock = WinsForms.DockStyle.Fill
        self.leftwin.Location = WinsDraw.Point(0,0)
        self.leftwin.PanelMinSize = int(self.Width/5 + 40)
        self.leftwin.Size = WinsDraw.Size(int(self.Width/5 + 50),
                                 int(self.Height))
        self.leftwin.ForeColor = Color.White #Black
        self.leftwin.BorderStyle = BorderStyle.FixedSingle
```

```
        self.rightwin = self.readerContainer.Panel2
        #self.rightwin.Dock = WinsForms.DockStyle.Fill
        self.rightwin.Location = WinsDraw.Point(0,0)
        self.rightwin.PanelMinSize = int(self.Width/2 + 100)
        self.rightwin.BackColor = Color.LightSlateGray
        self.rightwin.Size = WinsDraw.Size(int(self.Width * 3/4),
                                           int(self.Height))
        return True

    def InstanceRightPanel(self):
        self.rPanel = WinsForms.Panel()
        self.rPanel.Location = WinsDraw.Point(5, 5)
        #self.rPanel.Height = self.rightwin.Size.Height - 10
        #self.rPanel.Width = self.rightwin.Size.Width - 10
        self.rPanel.Size = WinsDraw.Size(int(self.Width * 3/4),
                                         int(self.Height))
        self.rPanel.BackColor = Color.Navy
        self.rPanel.ForeColor = Color.White
        self.rPanel.BorderStyle = BorderStyle.FixedSingle
        return True

    def InstanceLeftPanel(self):
        self.lPanel = WinsForms.Panel()
        self.lPanel.Location = WinsDraw.Point(5, 5)
        #self.lPanel.Height = self.leftwin.Size.Height - 10
        #self.lPanel.Width = self.leftwin.Size.Width - 10
        self.lPanel.Size = WinsDraw.Size(int(self.Width * 3/4),
                                         int(self.Height))
        self.lPanel.BackColor = Color.LightGray
        self.lPanel.ForeColor = Color.Navy
        self.lPanel.BorderStyle = BorderStyle.FixedSingle
```

```
            return True

        def GUIFormShow(self):
            self.Show()
            return True

      def __del__(self):
            pass
```

Indeed, the pair Container-objects are perceptible only because and after the implementation on the Class-derived form Object by the mounting relation

```
self.Controls.Add(self.readerContainer)
```

which in fact creates the Container pair panels on the Form suited for display accentuated by the Panels BackColors differences.

6.5 Developing .Net TabbedDialog

TabbedDialog is a reserved space on the Form in which texts are usually written either for identification or for giving instructive guidance or explanation of a particular action to the user. The TabbedDialog is particularly useful when screen space constraints exists and more information must be displayed than there is space to display; the TabbedDialog effective enables a sort of layering in which one set of displayable data are layered above another, one set at a time yet all accessible to the end-user.

As with all Windows objects the TabbedDialog is also of a nested class of the Form class and as such it is an inner class of the form class and so create-specified by the standard inner classes instantiation method, as per given in Section 4.6 is instruction:

```
self.tbcExample = WinsForms.TabControl()
```

effectively causing a kind of Control-object authorization to subsequently create on a Windows Form the fully coded a TabPages layering of a Form object suited for display in a GUI sets of Dialog mounted objects as shown in Fig.6.4, with a mere Panel object which is used to further contain and manage the layout of GUI that empowers the end-user of interaction with the GUI objects as well as

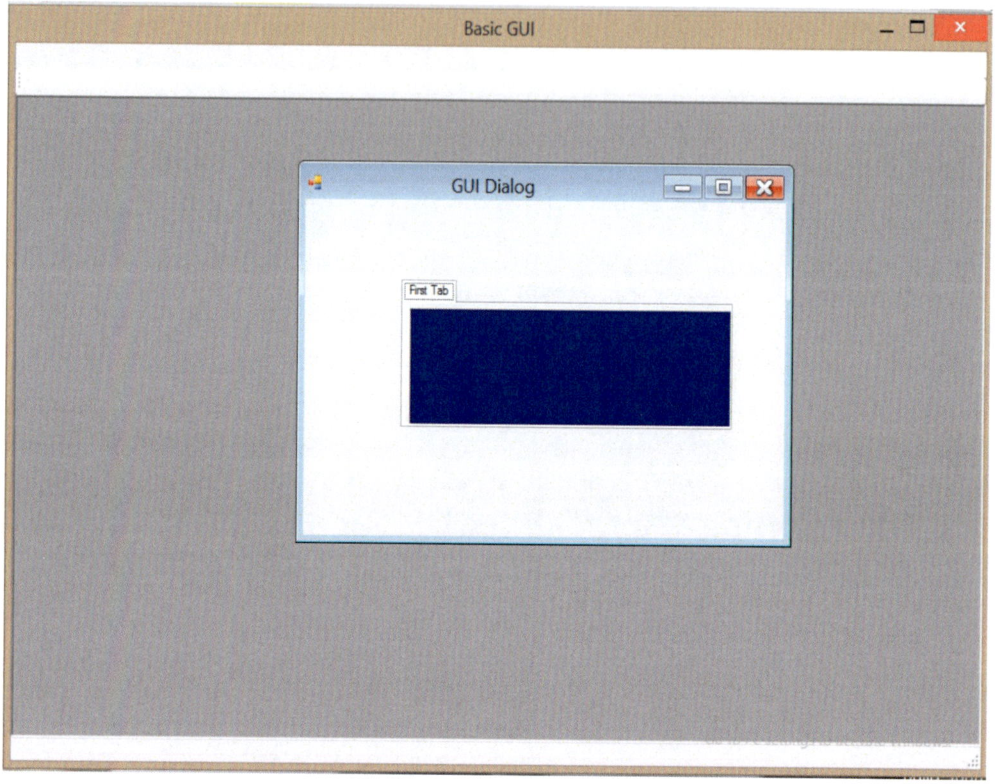

Figure 6.4: TabbedDialog Example pythonNet GUI

notification of tab access availability to other GUI objects sets.

The code for instantiation an object of the TabbedDialog class on the form is given in the full with respect to the common properties of form objects as listed in Section 4.7, while adapted for more specificity

```
import guipymode
from guipymod import *
```

```
import huwinsforms
from guiwinsforms import *
import formbase.guilayout
from formbase.guilayout import SetUpGUIForm

class ExampleTabbedDialog(Form)
    def __init__(self, parent):
        Form.__init__(self)
        self.MdiParent = parent
        self.parent = parent
        self.Text = "GUI Dialog"
        self.Height = self.parent.Height/2
        self.Width = self.parent.Width/2
        self.Location = WinsDraw.Point(int(self.parent.Width/4),
                              int(self.parent.Height/4))

        self.InstanceTabbedDialog()
        self.InstanceFirstPanel()

        self.wtcExampleVPage.Controls.Add(self.fPanel)
        self.wtcExample.Controls.Add(self.wtcExampleVPage)
        self.Controls.Add(self.wtcExample)
        self.ResumeLayout(False)

        self.GUIFormShow()

    def InstanceTabbedDialog(self):
        self.wtcExample = WinsForms.TabControl()
        self.wtcExample.SuspendLayout()
        self.wtcExample.Location = WinsDraw.Point(100,75)
        #self.wtcExample.Width = 500
```

```
        #self.wtcExample.Height = 500
        self.wtcExample.Size = WinsDraw.Size(
            int(self.Width * 3/4 - 40), int(self.Height/2 - 40))
        self.wtcExample.TabIndex = 0
        tpTitle = str("First Tab")
        self.wtcExampleVPage = WinsForms.TabPage(tpTitle)
        return True

    def InstanceFirstPanel(self):
        self.fPanel = WinsForms.Panel()
        self.fPanel.Location = WinsDraw.Point(5, 5)
        #self.fPanel.Height = self.Size.Height - 10
        #self.fPanel.Width = self.Size.Width - 10
        self.fPanel.Size = WinsDraw.Size(int(self.Width * 3/4),
                       int(self.Height))
        self.fPanel.BackColor = Color.Navy
        self.fPanel.ForeColor = Color.White
        self.fPanel.BorderStyle = BorderStyle.FixedSingle
        return True

    def GUIFormShow(self):
        self.Show()
        return True

    def __del__(self):
        pass
```

the object-specific codes for the TabbedDialog obj are the following and can be determined from examples of the assignments to these properties: Similarly the object specific methods of the TabbedDialog object are as follows:

```
self.wtcExample.SelectedTab
```

```
self.wtcExample.SelectedIndex
```

These object-specific properties acts such as enable the determination of the name of the tabpage that ot selected and that should be selected. By example the use of SelectedTab is as follows:

```
seltab = self.wtcExample.SelectedTab
```

will enable extracting the name from the Tab. The SelectedIndex enables the performing of a mirror task; say, a specific tab is desired to be selected, and the index of the tab is known, then it is a simple code to write as:

```
index = self.wtcExample.SelectedIndex
self.wtcExample.SelectedTab = index
```

with the result that the appropriate tab is selected and the content objects displayed for use by the end-user of the application.

Chapter 7

Contextual Logical Designs

7.1 Contextual Data Flows

The coding of the Windows Forms Objects as in Chapter ?? addressed mostly initialization dataflows in which the flow of data goes into an object from well-defined Data Structures. However, data flows from one object to another object during use involves a case of the transfer of data from data structure of source objects that may be incompatible with destination data structures required for entry into the destination objects. Of particular interest then is the coding of the data flow involving such data structures transformations that enables compatibility.

data structures transformations

Even then the start of the data flow process is the raising of Events and the coding of the Event handlers that enables the collecting of data from one object or sets of objects for input into another object or into the information logical design of the software. So then a of core consideration of contextual data flow is the the EventHandlers.

Collecting data from TextBoxes into other objects entails Object specific properties and methods invocations as may have been noticed except for the containers, data may be collected from direct entry by the end-user or from file or any other storage medium. Initialization data may yet be extracted from the specific objects and then transferred into another object for directive of subsequent data extraction.

The reformatting and structure-specific aggregations of data from GUi objects into specific Data Structures for input into other GUI objects undoubtedly should involve similar steps as the issue of data re-aggregation as with Databases data in the operations of ETL - Extract, Transform and Load. So some considerations for such data activities is anticipated to be performed in the contextual logical designs of the pythonNet GUIs.

The data operation as observed so far on in-transit data requires a level of conversance with the various operational python Data structures for effective pythonNet coding.

7.2 DataFlow Data-Structures

The python Data structures that are available for use in the flow of data between the pythonNet GUI objects are the List, Array, Tuple and Dictionary. These data structure while few compared to other languages are effectively used to construct other data structures of other languages as needed, and as such these Data Structures are sufficient for the ETL of data from one GUI object to another.

7.2.1 The List Data-Structure

The python List Data-Structure is the most used in python programming particularly useful as its structure and operations also form the foundations for two other Data Structures: Array and Tuple. The List Data structure is defined as a data container variable, say, dlist, having the form:

```
dlist = [x, y, z]
```

Notably, however, List are mutable: elements of a List Data Structure can be varied as a result of remote operations not directly performed on the list: An operation on a copy of the list changes the elements of the original list. The list is created in the memory by the declaration:

```
dlist = []
```

and the constructed into the form above containing the data set: x,y, z, as before, using the built function append in the insertion relations as follows:

```
dlist = []
dlist.append(x)
dlist.append(y)
dlist.append(z)
```

The elements of the List, as say, dlist, are read out as variable, v, by assignment of the indexed list data

```
v = dlist[i]
```

The python List is also dynamically created from file lines by one of the methods for reading text-files as shown: given a text file specified by the file path filepath, the following operations generates a list data

```
filepath
fl = open(filepath, '"r")
flist = fl.readlines()
```

with the output flist being a list, just as the list of dlist, and then the list element can be extracted as before.

Lists data are also dynamically produced from comma-delimited strings, "x, y, z"' with the use of another built function split in the following operations:

```
strd = "x, y, z"
strdlist = strd.split(",")
```

generation yet again a List data, strdlist, which are also all subject to the operation of of the List data Structure.

As previously observed the python List Data Structure is also the foundation for the Data Structures Array and the presentation of the array elicits the observations.

7.2.2 Python Array

Python Array is simply a reformed python List and is formed only from a python List, using the following declaration –using the same set of data, x, y, z as with the dlist to emphasize the construct :

```
darr = array([x, y, z])
```

with a parenthesis added as further bounding on the List Data Structure. As with the List data Structure, the elements of the array are read out by indexing the array data:

```
v = darr[i]
```

Also as previously observed the python List Data Structure also forms the foundation for the Tuple Data Structures and the presentation of the tuple elicits the observations.

7.2.3 Python Tuple

Python Tuple Data structure is simply a restatement of python List and is formed only with the same data sequence as of a python List, with the following declaration –using the same set of data, x, y, z as with the dlist to emphasize the construct :

```
dtuple = (x, y, z)
```

with a parenthesis replacing the square parenthesis as bounding on the similarly sequenced data of List Data Structure. The tuple structure can also be created in memory by declaring empty tuple to begin with as:

```
dtuple = ()
dtuple = tuple()
```

Once declared the tuple could be loaded with data fro use by the mere use of the addition operation:

```
dtuple = dtuple + (x, y, ..z)
dtuple += (x, y,..., z)
```

where the numbers in parenthesis is akin to the the same as used in the earlier method of creating the tuple from the list data, dlist.

However, tuples are immutable: elements of the tuple object from which a copy is made does not change when any element of the tuple-copy is changed. So it is preferably used to pass data from unction to function.

As with the List data Structure, the elements of the tuple also are read out by indexing the tuple data:

```
v = dtuple[i]
```

yet the python Tuple Data Structure contrasts significantly with a List Data Structure in the sense that while the elements of a List can be varied from operations, the data-elements of a Tuple Data Structure can not be changed. So this Data structure is very effective for data transfers that must ensure constancy of the constituent variables.

7.2.4 Python Dictionary

The python Dictionary Data structure is distinctively defined relative to the List data structure. The same set of data, "x, y, z" could also have been stored in a python dictionary, ddict, for subsequent transfer or insert into another object. Such use may be necessitated by the need for specificity of order or association in the use of the data set. A python dictionary by definition has the structure:

```
ddict = {"fn": "x", "sn": "y", "tn": "z"}
```

where the fx: x = n, n =1,2,3; are order specifier such as first number and so on; however, those additional information are actually termed as keys while the data being actually transferred are termed values. A python dictionary as such is a key-value pair mechanism of data containment.

The python dictionary is statically created by the following set of codes: the first code in essentially creates the empty dictionary, and the subsequent lines of codes makes the assignment of the data into the specific key

```
ddict = dict()
ddict["fn"] = "x"
ddict["sn"] = "y"
ddict["tn"] = "z"
```

The same dictionary data could also have been created by insertion of elements of a List data by indexing the List data, and is elicited with the use of the List data, dlist, created above using indexing:

```
ddict["fn"] = dlist[1]
ddict["sn"] = dlist[2]
ddict["tn"] = dlist[3]
```

and the data elements or values of the key-value pair of the dictionary read out by specification of the relevant key, akin to the use of index in the case of List Data structure, as follows:

```
v = ddict[key]
```

Alternatively a built-in function get can also be used to retrieve the value-data associated with a specific key as follows:

```
v = ddict.get(key)
```

Of course, in both cases it s presumed that the key is known as is with the case of fn, sn, tn one of which must be explicitly stated in place of key.

In those cases where the entire data set need to be extracted but the key need not be then the extraction can be effected by looping through the keys as follows:

```
for key in ddict.keys()
    v = ddict[key]
```

Alternatively using the built in get function by the following code:

```
for key in ddict.keys()
    v = ddict.get(key)
```

Then, of course, for those cases for which both the key and value data are wanted then the iteration could take the form:

```
for key in ddict.keys()
    kd = (str(key))
    vd = ddict.[key])
```

while the iteration extracts the individual key-value pairs the set may be stored in a list data if so desired by the developer.

For completion the initialization of the dictionary data, say, ddict, could very well have been performed with the following code, involving the use of array:

```
karr = array(["fn", "sn", "tn"])
arrcnt = len(karr)
for i in range(0, arrcnt):
    ddict[karr[i]] = dlist[i]
```

Similarly the array data could very well have been substituted for with a tuple data and the result will be the same, and the dictionary data, ddict still initialized.

7.3 ComboBox Data Exchanging

ComboBox as a data container is often filled with data and so the process of inserting data dynamically into a ComboBox object from either a List data or Dictionary data is of some import in the coding of pythonNet GUI.

7.3.1 List Data Source

Inserting data from a python dictionary data into a ComboBox object also uses the dictionary data extraction methods for dictionaries and is used with ComboBox object as follows:

```
cbbExample = WinsForms.ComboBox()
cbbExample.BeginUpdate()
lstcnt = len(dlist)
for i in range(0, lstcnt)
    cbbExample.Items.Add(dlist[i])
cbbExample.EndUpdate()
```

or alternatively the data inserts could be performed using the built-in insert function as follows:

```
cbbExample = WinsForms.ComboBox()
lstcnt = len(dlist)
for i in range(0, lstcnt)
    cbbExample.Items.insert(i, dlist[i])
```

using the looping index to also index the data variable within the ComboBox object. of course, the index in the insert function need not be exactly the same i, a different numerical evaluated by any suitable method could have been used.However, the advantage of using the insert function is that the setting of the visible enrty of the ComboBox gets to e easily done with the following codes:

```
cbbExample.SelectedIndex = j
```

where the value of j is the index of choice associated with the preferred visible data of the ComboBox object.

Then, of course, where no explicit indexing of the data items in the ComboBox object is unnecessary then then data insertion into the CombBox object could have been accomplished with as follows:

```
for litm in dlist
    cbbExample.Items.Add(litm)
```

The data entry into a ComboBox from the list data and by relation, also from array and tuples can be accomplished as by the above methods.

7.3.2 Dictionary Data Source

Inserting data from a python dictionary data into a ComboBox object also uses the dictionary data extraction methods for dictionaries and is used with ComboBox object as follows:

```
cbbExample.BeginUpdate()
for key in ddict.keys()
    cbbExample.Items.Add(ddict[key])
cbbExample.EndUpdate()
```

or alternatively the built-in get function is used as follows:

```
cbbExample.BeginUpdate()
for key in ddict.keys()
    cbbExample.Items.Add(ddict.get(key))
cbbExample.EndUpdate()
```

The data entry into a ComboBox from the dictionary data are generally accomplished as by the above methods.

7.3.3 ComboBox Data Retrieval

The flow of data out of a ComboBox object, while using the object-specific Events raised for the transfer as discussed under the logical design of the object in Section 5.8, requires reasoned use of the Events of the object raised as a result of the action of the user. The reasoned use of these Events and the proper coding of the related EventHandler is significant in the proper working of Form objects. The primary concern is that the data will flow from any of python Data Structures and the collection each type is critical.

The Events specific to ComboBox ithat are of particular interest in enablng data retrieval from a ComboBox object are two: Click and AfterSelection

- Click

- AfterSelection

Typically a click of a ComboBox has the object of selecting a content of the dropdown box; however, the data immediately visible and available as the one prevailing prior to the Click Event, yet under certain conditions of the click, is selected. So, the Click Event naturally recognizes the visible entry of the ComboBox as the new selection, although sometimes it may be delayed long enough as to recognize the selection that becomes visible after the entry-selecting click. This behaviour therefore is non-repeatable and needs be avoided.

On the other has the AfterSelection always recognize the visible item after the DropDown has pulled back up. So then the reasoned use of these Events and the proper coding of the related EventHandler suggests the use of the AfterSelection Event as it captures the interactions with both Events, and the coding could be as follows:

```
cbbexample.AfterSelection += self.AfterSelectionProcess

def AfterSelectionProcess(self), sender, eventargs):
    if sender ++ cbbExample:
        vItem = self.cbbExample.SelectedItem
    return True
```

making the data now borne by vItem available for use in any other sections of the codes, concluding the data flow need as applicable to ComboBoxes.

7.4 ListBox Data Exchanging

ListBox as a data container is often filled with data and so the process of inserting data dynamically into a ListBox object from either a List data or Dictionary data is of some import in the coding of pythonNet GUI.

7.4.1 List Data Source

The Events raised by ListBox are akin to those of the ComboBox and so the codes for the ComboBox could be readily adapted for use as well for the ListBox object. Effectively inserting data from a python List data into a ListBox object also uses the list data extraction methods for list data and is used with ListBox object as follows:

```
ltbExample = WinsForms.ListBox()
ltbExample.BeginUpdate()
lstcnt = len(dlist)
for i in range(0, lstcnt)
    ltbExample.Items.Add(dlist[i])
ltbExample.BeginUpdate()
```

then then data insertion into the ListBox object could have been accomplished with as follows:

```
ltbExample = WinsForms.ListBox()
ltbExample.BeginUpdate()
for litm in dlist
    ltbExample.Items.Add(litm)
ltbExample.BeginUpdate()
```

7.4.2 Dictionary Data Source

Inserting data from a python dictionary data into a ListBox object also uses the dictionary data extraction methods for dictionaries and is used with ListBox object as follows:

```
for key in ddict.keys()
    ltbExample.Items.Add(ddict[key])
```

7.4.3 ListBox Data Retrieval

The flow of data out of a ListBox object, while using the object-specific Events raised for the transfer as discussed under the logical design of the object in Section 5.9, requires reasoned use of the Events of the object raised as a result of the action of the user. The reasoned use of these Events and the proper coding of the related EventHandler is significant in the proper working of Form objects. The primary concern is that the data will flow from any of python Data Structures and the collection each type is critical.

For instance the Events that are custom to ListBox include

- Click

The reasoned use of these Events and the proper coding of the related EventHandler is significant in the proper working of Form objects.

```
ltbExample.Click += self.ltbExampleClick

def ltbExampleClick(self, sender, eventargs):
    if sender ++ self.ltbExample:
        vItem = ltbExample.SelectedItem
    return True
```

Typically a click of a ListBox has the object of selecting an item of the listbox over which the mouse is clicked. So then Data Flow due to data retrieval from ListBox object can be effected with a mere click over the data.

7.5 ListView Data Exchanging

ListView as a data container displays data in the form of a mathematical matrix elements layout: Rows ans columns. In c# the object is often filled with data by mechanism that may be described as columnar or column-wise, while in pythonNet the object if filled with data by mechanism that may be deemed as row-wise. In each case the the process of inserting data dynamically into a ListView object must be based on the Data structures that align with the data insert algorithm. The data structures from which the data is retrieved to fill the Listview must allow the specific dataset suited for the algorithm.

In particular, each data set must be a key-value pair in which the key is an dataset counter index while the value must be a data container of length the same as the number of columns as the Listview. In effect, the data structure for filling each row must a python Dictionary data of the form:of one these:

```
vd = {key:[x1, x2, x3, ... xn, ...xL]}
vd = {key:([x1, x2, x3, ... xn, ...xL])}
vd = {key:(x1, x2, x3, ... xn, ...xL)}
```

the first is of a List Data Structure, the second is of the Array Data Structure and the third is of the for of a Tuple Data Structure; and this requirement is imposed by the need for implementing the following data insertion into Listview object method:

```
self.ltvExampleItem = WinsForms.ListViewItem(key,0)
self.ltvExampleItem.SubItems.Add(x1))
self.ltvExampleItem.SubItems.Add(x2))
self.ltvExampleItem.SubItems.Add(x3))
.
.
.
self.ltvExampleItem.SubItems.Add(xn))
.
.
.
self.ltvExampleItem.SubItems.Add(xL))
self.ltvExample.Items.Add(self.ltvExampleItem)
```

as previously shown in section 5.11. Of course, the entry in the declaration invocation can always start with the first value element and not the key.

Obvious from this requirement elicitation is that the actual data type with which to start the data insert is a complex data structure of any of the python Data structures such as one of the nested data structure data types:

```
List data-type:=  [{key:[x1, x2, x3, ... xn, ...xL]},
                   {key:[x1, x2, x3, ... xn, ...xL]}, ...]
Array data-type:=  ([{key:[x1, x2, x3, ... xn, ...xL]},
                    {key:[x1, x2, x3, ... xn, ...xL]}, ...])
Tuple data-type:=  ({key:[x1, x2, x3, ... xn, ...xL]},
                    {key:[x1, x2, x3, ... xn, ...xL]}, ...)
Dictionary data-type:=  {{key:[x1, x2, x3, ... xn, ...xL]},
                         {key:[x1, x2, x3, ... xn, ...xL]}, ...}
```

These then are the data structures for which codes must be developed for insert into ListView objects.

Yet because the data retrieval methods are essentially the same for the List, Array and Tuple data Structures only the List data structure type is coded presently in further eliciting of the rationale.

7.5.1 List Data Source

Inserting data from a python List data into a Listview object also uses the list data extraction methods for list data and is used with Listiew object as follows: First the flow from a complex list object is considered:

```
ltvExample = WinsForms.Listview()
dsrc =  [{key1:[x1, x2, x3, ... xn, ...xL]},
         {key2:[y1, y2, y3, ... yn, ...yL]}, ...]
lstcnt = len(dlist)
ltvExamle.BeginUpdate()
for i in range(0, lstcnt):
```

```
        delem = dsrc[i]
        for key in delem.keys()
            ltvExampleItem = WinsForms.ListviewItem(key, 0)
            dvalset = delem[key]
            for item in dvalset:
                ltvExampleItems.SubItemAdd(item)
            self.ltvExample.Items.Add(self.ltvExampleItem)
    ltvExamle.EndUpdate()
```

A necessary observation here is that the data is entered into the Listview object one row at a time, while in the case of C# each column corresponding to the columns of the listview is first filled up and then all the columns simultaneously inserted into the Listview object with a function, "AddRange()"'

7.5.2 Dictionary Data Source

Inserting data from a python dictionary data into a Listview object also uses the dictionary data extraction methods for dictionaries and is used with Listview object as follows:

```
ltvExample = WinsForms.Listview()
dsrc =  {{sk1:{key1:[x1, x2, x3, ... xn, ...xL]}},
            {sk2:{key2:[y1, y2, y3, ... yn, ...yL]}}, ...}
ltvExamle.BeginUpdate()
for key in dsrc:
    delem = dsrc[key]
    for key in delem.keys()
        ltvExampleItem = WinsForms.ListviewItem(key, 0)
        dvalset = delem[key]
        for item in dvalset:
            ltvExampleItems.SubItemAdd(item)
        self.ltvExample.Items.Add(self.ltvExampleItem)
ltvExamle.EndUpdate()
```

7.5.3 ListView Data Retrieval

The flow of data out of a Listview object, while using the object-specific Events raised for the transfer as discussed under the logical design of the object in Section 5.11, requires reasoned use of the Events of the object raised as a result of the action of the user. The reasoned use of these Events and the proper coding of the related EventHandler is significant in the proper working of Form objects. The primary concern is that the data will flow from any of python Data Structures and the collection each type is critical.

The significant Event of a Listview is the Click Event

- Click

The click of a Listview object by default returns the data of the first column, but will return all data across the row entry when the MultiSelect property is set like

```
self.ltvEsample.MultiSelect = True
```

and in such cases the data across the row are collected; however, as the row data across only represents an item of the feed data construct for inserting data into the object, the data structure as obtained can be viewed as of the form:

```
{key:[x1, x2, x3, ... xn, ...xL]}
```

So then the proper coding of the related EventHandler for retrieval of data from a Listview object should aim to construct data of the same structure as given.

```
ltvExample.Click += self.ltvExampleClick

def ltvExampleClick(self, sender, eventargs):
    lvlist = []
    lvdict = dict()
    if sender == self.ltvExample:
        lvtlist = ltvExample.SelectedItems
        lvtdict[key] = lvtlist
    return True
```

Typically a click of a Listview object selects a row of the Listview data over which the mouse is clicked, and makes the SelectedItems into a list.

7.6 TreeView Data Exchanging

Treeview as a data container displays data in the form of a tree and so the process of inserting data dynamically into a TreeView object must be based on the Data structures that align with the data insert algorithm. The data structures from which the data is retrieved to fill the TreeView must allow the specific dataset suited for the algorithm.

In particular, as previously shown in section 5.10, implementing the following data insertion into Treeview object method:

```
self.trvExample.BeginUpdate()
parNode = WinsForms.TreeNode()
parNode.Name = str(key)
parNode.Text = str(key)
#parNode.Index = i
parNode.Tag = i
self.trvExample.Nodes.Add(parNode)
br = vd[key]
for item in br:
    childNode = WinsForms.TreeNode()
    childNode.Text = territory
    childNode.Name = territory
    nodeindex += 1
    j += 1
    childNode.Tag = nodeindex
    self.trvExample.Nodes[i].Nodes.Add(childNode)
self.trvExample.EndUpdate()
self.Conrtrols.Add(self.ltvExample)
```

Of course, the entry in the declaration invocation can always start with the first value element and not the key.

Obviously, each data set must be a key-value pair in which the key is defines the parent-node while the value dataset define the child-nodes, the data cardinality of which may or may not be the same. The cardinality variation is a significant feature in that it precludes the need for use of Table Data Structure such as proffered by Listview, as implemented prior.

In effect, the data structure for filling each row must a python Dictionary data of the form:of one these:

```
vd = {key:[x1, x2, x3, ... xn, ...xL]}
vd = {key:([x1, x2, x3, ... xn, ...xL])}
vd = {key:(x1, x2, x3, ... xn, ...xL)}
```

the first is of a List Data Structure, the second is of the Array Data Structure and the third is of the for of a Tuple Data Structure.

Obvious from this requirement elicitation is that the actual data type with which to start the data insert is a c0omplex data structure of any of the python Data structures such as one of the nested data structure data types:

```
List data-type:=  [{key:[x1, x2, x3, ... xn, ...xL]},
                    {key:[x1, x2, x3, ... xn, ...xL]}, ...]
Array data-type:=  ([{key:[x1, x2, x3, ... xn, ...xL]},
                    {key:[x1, x2, x3, ... xn, ...xL]}, ...])
Tuple data-type:=  ({key:[x1, x2, x3, ... xn, ...xL]},
                    {key:[x1, x2, x3, ... xn, ...xL]}, ...)
Dictionary data-type:=  {{key:[x1, x2, x3, ... xn, ...xL]},
                     {key:[x1, x2, x3, ... xn, ...xL]}, ...}
```

These then are a set of data structures for which codes must be developed for insert into Treeview objects.

Yet these data Structures are simply the first tier forms for use; after all, because a tree has branches and the branches have branches, and so on till the leaves prevail, the data Structures can be whole more nested, with nested dictionary the elements of the list as per the example:

```
vd = [{x1:[x11, x12, x13, ... x1k, ...x1N]},
           {x2:[x21, x22, x23, ... x2k, ...x2N]},
           {x2:[x21, x22, x23, ... x2k, ...x2N]},
           .
           .
           .
           {xk:[xk1, xk2, xk3, ... xkk, ...xkN]},
           .
           .
           .
            {xM:[xM1, xM2, xM3, ... xMk, ...xMN]}]
```

and there can even more complex Data structures than the above and more so given that even the inner nested List data could very well have been arrays, tuples and even dictionaries. Even worsen the nested list could have been a mix of arrays, tuples and dictionaries. By similar reasoning even the starting List defining vd could have also been any of the data structures of array, tuple and dictionary. In effect, the implementing of a Treeview from python data structures can be messy and even harrowing under certain conditions.

In any event for the purposes of lucid elicitation of the underlying concepts, the illustrations of data Flow to a Treeview are restricted to the the first tier Data structures as first crafted above.. Further, because the data retrieval methods are essentially the same for the List, Array and Tuple data Structures only the List data structure type is coded presently in further eliciting of the concepts.

7.6.1 List Data Source

The reasoned use of these Events and the proper coding of the related EventHandler is significant in the proper working of Form objects. The primary concern is that the data will flow from any of python Data Structures and the collection each type is critical. First the flow from a complex list object is considered:

```
ltvExample = WinsForms.Listview()
dsrc =  [{key1:[x1, x2, x3, ... xn, ...xL]},
           {key2:[y1, y2, y3, ... yn, ...yL]}, ...]
lstcnt = len(dlist)
self.trvExample.BeginUpdate()
for i in range(0, lstcnt):
    delem = dsrc[i]
    for key in delem.keys()
        parNode = WinsForms.TreeNode()
        parNode.Name = str(key)
        parNode.Text = str(key)
        #parNode.Index = i
        parNode.Tag = i
        self.trvExample.Nodes.Add(parNode)
        br = delem[key]
        for item in br:
            childNode = WinsForms.TreeNode()
            childNode.Text = territory
            childNode.Name = territory
            #print(r"childNode.Name %10s"%(childNode.Name))
            nodeindex += 1
            j += 1
            childNode.Tag = nodeindex
            self.trvExample.Nodes[i].Nodes.Add(childNode)
self.trvExample.EndUpdate()
self.Conrtrols.Add(self.ltvExample)
ltvExamle.EndUpdate()
```

7.6.2 Dictionary Data Source

Inserting data from a python dictionary data into a Treeview object also uses the dictionary data extraction methods for dictionaries and is used with Treeview object as follows:

```
ltvExample = WinsForms.Listview()
dsrc =  {{sk1:{key1:[x1, x2, x3, ... xn, ...xL]}},
            {sk2:{key2:[y1, y2, y3, ... yn, ...yL]}}, ...}
ltvExamle.BeginUpdate()
for key in dsrc:
    delem = dsrc[key]
    for key in delem.keys()
        parNode = WinsForms.TreeNode()
        parNode.Name = str(key)
        parNode.Text = str(key)
        #parNode.Index = i
        parNode.Tag = i
        self.trvExample.Nodes.Add(parNode)
        dvalset = delem[key]
        for item in br:
            childNode = WinsForms.TreeNode()
            childNode.Text = territory
            childNode.Name = territory
            #print(r"childNode.Name %10s"%(childNode.Name))
            nodeindex += 1
            j += 1
            childNode.Tag = nodeindex
            self.trvExample.Nodes[i].Nodes.Add(childNode)
self.trvExample.EndUpdate()
self.Conrtrols.Add(self.ltvExample)
ltvExamle.EndUpdate()
```

7.6.3 TreeView Data Retrieval

The flow of data out of a TreeView object, while using the object-specific Events raised for the transfer as discussed under the logical design of the object in Section 5.10, requires reasoned use of the Events of the object raised as a result of the action of the user. The reasoned use of these Events and the proper coding of the related EventHandler is significant in the proper working of Form objects. The primary concern is that the data will flow from any of python Data Structures and the collection each type is critical.

For a typical Treeview object the mostly useful Events that could be raised for manipulating the contained data are these two:

- Click
- AfterSelection

So for retrieval of data from a Treeview object one or both of these two events must be handled by the developer with applicable codes. Then, the reasoned use of these Events and the proper coding of the related EventHandler is significant in the proper working of Form objects; and illustrative codings are presented.

Typically a click of a TreeView has the object of selecting a node of the expanded node-branch, and ordinarily, the Click Event recognizes the visible Clicked Node of the Treeview as the new selection. However, sometimes it may default to recognizing the last parent node as the new selection. This behaviour therefore is non-repeatable and needs be avoided.

On the other has the AfterSelection always recognize the visible item after the DropDown has pulled back up which suggests the use of the AfterSelection. At a start an EventHandler is defined for the Event when the user should happen to raise the Event during use:

```
self.trvExample.AfterSelection += self.AfterSelectionProcess
```

Following defining the EventHandler for the Event, the EventHandler is then fleshed out with codes as :

```
def AfterSelectionProcess(self, sender, eventargs):
    if sender == self.trvExample:
        selNode = self.trvExample.SelectedNode
    return True
```

making the data now borne by selNode available to retrieve parent and or child nodes for use in any other sections of the codes, concluding the data flow need as applicable to Treeview.

Chapter 8

Applications Main Windows

8.1 Basic Architecture

As has been adopted as a standard the minimal GUI for an application consists of a set of menus": File Edit, View, Configure Help, and Windows; a set of Toolbars: Open, Copy, Paste, etc, a vertical scroll bar, and a Statusbar. In all then the the Basic GUI for an application before specialization to the use-specification of the application has the same form as Fig. 8.1. This then is the default or minimal main window for all application and should be produced in pythonnet application as the start-off base design.

In the rest of this chapter then the coding for developing the minimal main window for application will be developed continually for the developer to evolve with the thinking behind the development

8.2 Document Interfaces - The Basic GUI

The typical code for the Basic GUI as given in Fig. 8.1 consists of the template code as given in Chapter **??** but modified to include the codes for the default features of Menus, Toolbars, and Statusbar within the designated sections as shown:

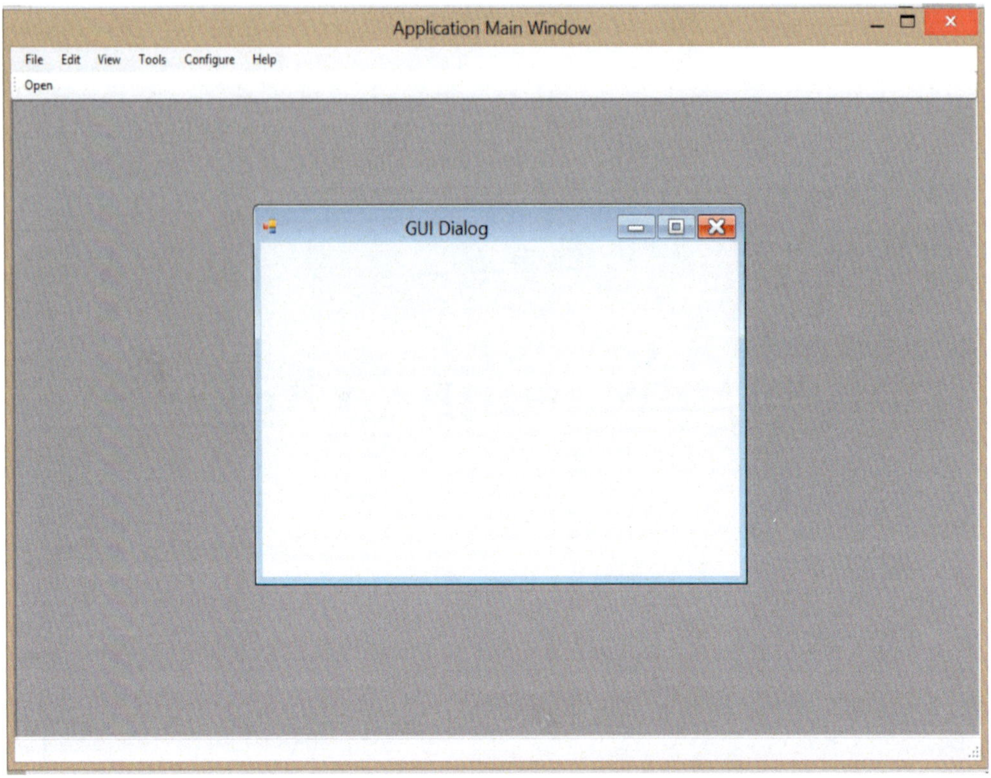

Figure 8.1: Applications GUI Main Window

```
import guipymode
from guipymod import *
import huwinsforms
from guiwinsforms import *
import formbase.guilayout
from formbase.guilayout import SetUpGUIForm

class pyNetBasicProgram(Form):
    def __init__(self):
        self.SetUpGUIForm(self)
```

```
        \textit{codes for standard Windows functionality}

        self.GUIFormShow()

        \textit{Codes for GUI Dialog objects functionality}

    def SelfShow((self):
        self.Show()
        return True

    def __del__(self):
        pass
```

For specific GUI Dialog functionality, the codes developed under the Chapters 4, 5 and 7 are included in the sections designated and in whatever arrangement the developer deems necessary to meet the goals of the development.

So the specialization of the Basic GUI Interface codes into the Documents GUI code entails the inclusion of codes for each of the functionality which are developed presently.

8.3 Menu Development

The menus items collection that form the standard items-set for every document are the File. Edit, View, Tools of Configure, and Help.

The development of any one first menu item starts with the creating of a menu strip or panel, on which each item is defined. The menu strip if created on the invoking an instance of the .Net 5 Menu Strip object by the Constructor, MenuStrip() and then attaching that the Form MainMenuStrip as follows:

```
self.tscontner = ToolStripContainer()
self.tspTop = ToolStripPanel()
self.tspTop.Dock = WinsForms.DockStyle.Top
```

```
self.mstrip = WinsForms.MenuStrip()
self.mstrip.Dock = WinsForms.DockStyle.Top
self.MainMenuStrip = self.mstrip
self.Controls.Add(self.mstrip)
```

The result of the addition of the menu strip, mstrip, is this neat white band directly below the upper bound of the basic interface as is now show in the Fig. ?? as compared to Fig.1.1

However the File menu that shows on the menu strip which enables the end-user to interact with the application, has to be created and an Event on the specific menu raised: the Event method must be invoked in reaction to the Event. The first task of creating a menu is straightforward and is created with the ToolStripMenuItem method with the invocation such as:

```
self.tscontner = ToolStripContainer()
self.tspTop = ToolStripPanel()
self.tspTop.Dock = WinsForms.DockStyle.Top
self.mstrip = WinsForms.MenuStrip()
self.mstrip.Dock = WinsForms.DockStyle.Top
self.menu = ToolStripMenuItem(MenuName)
self.mstrip.items.Add(self.menu)
self.mstrip.MdiWindowListItem = self.menu
self.MainMenuStrip = self.mstrip
self.Controls.Add(self.mstrip)
```

which creates a menu of name *self.menu.*

The next step is to attach an EventHandler to this menu so that, say, on clicking the menu the application does something. There are two approaches to addressing the Event that is raised: Use of EventHandler or CancelEventHandler to cancel the Event as raised by the system. The coding of a simple EventHandler entails the raising of the event by associating the Event method to an EventHandler as given in the code snippet:

```
self.tscontner = ToolStripContainer()
self.tspTop = ToolStripPanel()
self.tspTop.Dock = WinsForms.DockStyle.Top
self.mstrip = WinsForms.MenuStrip()
self.mstrip.Dock = WinsForms.DockStyle.Top
self.menu = ToolStripMenuItem(MenuName)
self.menu.Event += EventHandler(self.EventMethod)
self.mstrip.MdiWindowListItem = self.menu
self.mstrip.items.Add(self.menu)
self.MainMenuStrip = self.mstrip
self.Controls.Add(self.mstrip)
```

and now whenever the end-user does whatever is constitute the Event as defined in association, the EventHandler gets invoked, and the code instructions given as the body of the EventHandler is executed.

The second approach uses a CanceEventHandler which more instinctive is understood as Event-cancel handler meaning a code to can the Event raised by the user action, and is coded as

```
self.tscontner = ToolStripContainer()
self.tspTop = ToolStripPanel()
self.tspTop.Dock = WinsForms.DockStyle.Top
self.mstrip = WinsForms.MenuStrip()
self.mstrip.Dock = WinsForms.DockStyle.Top
self.menu = ToolStripMenuItem(MenuName)
self.menu.Event = System.ComponentModel.CancelEventHandler(
                                                  self.EventMethod)
self.mstrip.MdiWindowListItem = self.menu
self.mstrip.items.Add(self.menu)
self.MainMenuStrip = self.mstrip
self.Controls.Add(self.mstrip)
```

then, of course, the EventMethod is then coded as with the structure:

```
def EventMethod(self, sender, evertargs):
    if sender == self.menu:
        \textit{---codes here---}
    return True
```

the code above is a rather simplified form, depending on the Event raised the coding may get quite involved and be as long as necessary. By this format then the essential menus of an application made be added as such as File menu, as is the case next.

Noteably though, when the menu is not terminal but rather has submenu or submenus then the Event association and the corresponding EventHandlers are coded to the submenus instead of to the menu proper. Such a situation for a submenu aptly called *submenu* is given in this code snippet:

```
self.tscontner = ToolStripContainer()
self.tspTop = ToolStripPanel()
self.tspTop.Dock = WinsForms.DockStyle.Top
self.mstrip = WinsForms.MenuStrip()
self.menu.Dock = WinsForms.DockStyle.Top
self.menu = ToolStripMenuItem(MenuName)
self.submenu = ToolStripMenuItem(submenuName)
self.submenu.EventMethod += self.EventHandler
self.menu.DropDownItems.Add(self.submenu)
self.mstrip.MdiWindowListItem = self.menu
self.mstrip.items.Add(menu)
\emph{---optional more menus---}
self.MainMenuStrip = self.mstrip
self.Controls.Add(self.mstrip)
```

The general guidance then for coding a menus and submenus in a pythonNet GUI is as given so far.The approach presently is applied to the standard menus set of the Main Window in further elicitation by the specialization of the guidance.

8.3.1 File

The File menu is a means of granting access to the end-user to find files on the hard drive without actually leaving the application. The menus enables files on had drives to get loaded into the applications designated space. So the the menu has become one of the standard menus that must be included in all applications that involves access to materials on the Local Hard Drive.

The essential codes for creating a File menu is as given in the follow code:

```
import guipymode
from guipymod import *
import huwinsforms
from guiwinsforms import *
import formbase.guilayout
from formbase.guilayout import SetUpGUIForm

class pyNetBasicProgram(Form):
    def __init__(self):
        self.SetUpGUIForm(self)

        self.tscontner = ToolStripContainer()
        self.tspTop = ToolStripPanel()
        self.tspTop.Dock = WinsForms.DockStyle.Top
        self.filemenu = ToolStripMenuItem("File")
        self.fileOpenmenu = ToolStripMenuItem("Open")
        self.fileOpenmenu.Click +- self.FileOpenClick
        self.filemenu.DropDownItems.Add(self.fileOpenmenu)
        self.mstrip.MdiWindowListItem = self.filemenu
        self.mstrip.items.Add(self.filemenu)
        \emph{---optional more menus---}
        self.MainMenuStrip = self.mstrip
        self.Controls.Add(self.mstrip)
```

```
    def SelfShow((self):
        self.Show()
        return True

    def __del__(self):
        pass
```

The submenu state of the Open menu as created is as shown in Fig. 8.2 illustrates the dropdown property of menus as examplified by the File menu However, the

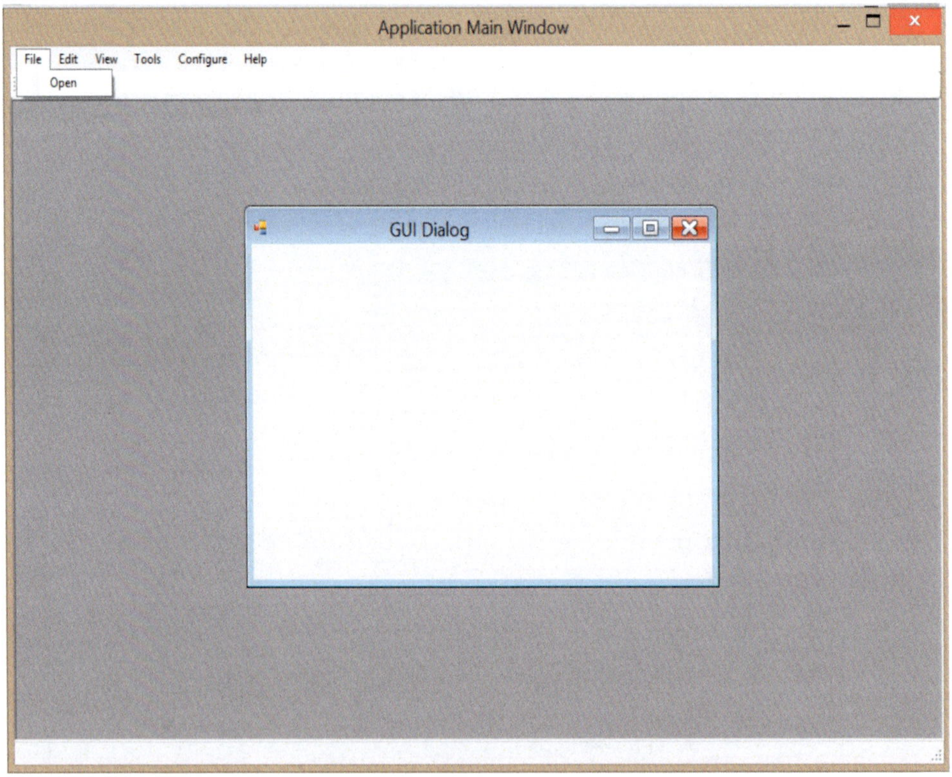

Figure 8.2: File Open pythonNet GUI

EventHandler of the File access menus always opens up a FileOpendialog interface from which files in the hard drive are selected to be opened. The Eventhandler, *self.fileOpenmenu* as such must invoke to have created an instance of the OpenFile Dialog interface, by this code:

```
def FileOpenmenu(self, sender, eventargs):
    self.fod = self.MyFileOpenDialog()
    return True
```

It should be noted that although the class derives from a Forms object, the invocation of instantiation only follows the argument requirements for the function: def __init__(self).

Coding the OpebFileDialog interface however is somewhat involved and in some cases even challenging. Indeed, the OpenFiledialog object has to be coded to enable the file search.While there is more than one approach to code this dialog, the simplest is to derive the dialog from the system design as here:

```
class MyFileOpenDialog(WinsForms.OpenFileDialog):
    def __init__(self, parent, caller):
        Form.__init__(self)
        self.ClientSize = WinsDraw.Size(460, 400)
        self.ScreenSize = WinsDraw.Size(460, 300)
        self.parent = parent
        self.caller = caller
        self.Width = 460 #parent.Width/2 - 20.00
        self.Height = 400 #parent.Height/2 *2.5
        self.Text = "Browse Text Files"
        #begin OpenFiledialog control contruction
        self.InitialDirectory - str("C:/myscripts/")
        self.RestoreDirectory = True
        #self.Title = "Click Data File"
        self.DefaultExt = 'txt'
        self.Filter = "Excel Files(*.txt)|*.txt"
        self.FilterIndex = 2
        self.CheckFileExists = True
```

```
        self.CheckPathExists = True
        self.GenerateMember = True
        #self.Modifiers = Private
        self.Filename = "OpenDFileDialog1"
        self.ReadOnlyChecked = True
        self.ShowReadOnly = True
        self.ConstructDialogUI()
        self.ShowDialog()

    def ConstructDialogUI(self):
        \emph{custom dialog design codes go here---}
        return True

   def __del__(self):
        pass
```

A typical OpenFile Dialog interface when instantiated with a click on the File submenu *Open* has the appearance as shown in Fig. 8.2 where the user then selects a directory from the ComboBox at the top and then a file at the middle section listing of files, the ListBox interface.

8.3.2 Edit

The Edit menu is a means of granting access to the end-user to find files on the hard drive without actually leaving the application. The menus enables files on had drives to get loaded into the applications designated space. So the the menu has become one of the standard menus that must be included in all applications that involves access to materials on the Local Hard Drive.

The codes for creating an Edit menu with the submenu, Copy, follows the same structure as earlier developed for the File menu and is as given in the follow code:

```
self.edEditmenu = ToolStripMenuItem("Edit")
```

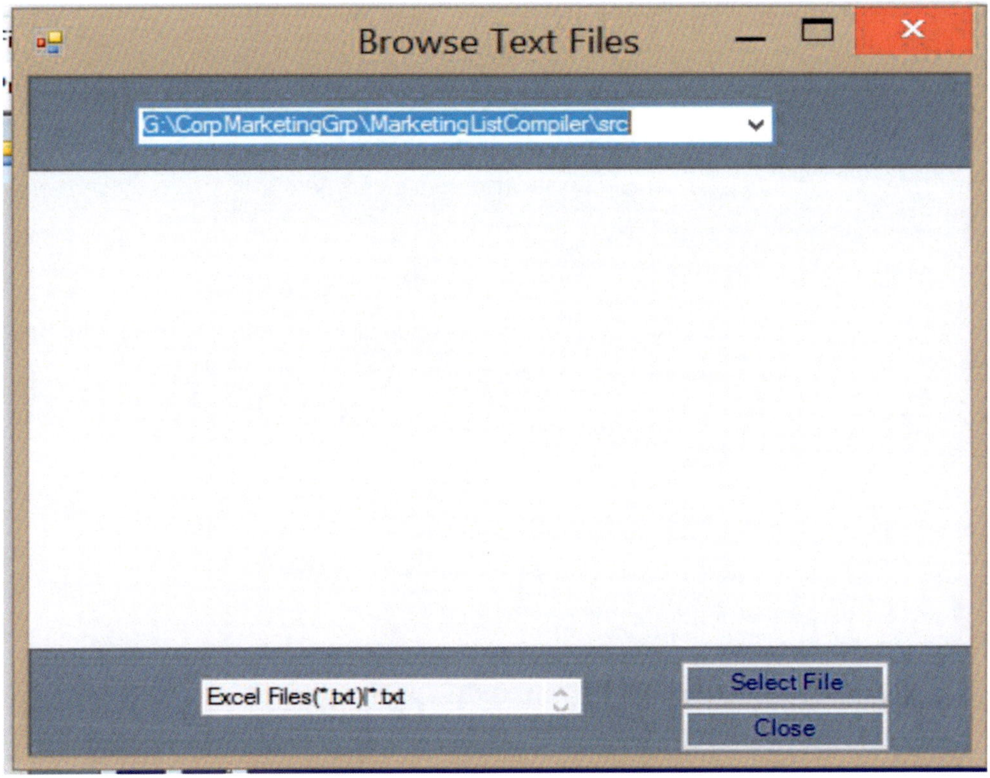

Figure 8.3: A Basic pythonNet GUI

```
self.editCopymenu = ToolStripMenuItem("Copy")
self.editCopymenu.Click += EventHandler(self.EditCopy)
self.edEditmenu.DropDownItems.Add(self.editCopymenu)
self.mstrip.MdiWindowListItem = self.edEditmenu
self.mstrip.items.Add(self.edEditmenu)
```

Naturally this set of codes is inserted, relative to the File menu codes; in the section marked for "'optional menus codes" so that the last two codelines continue to apply to these codes as well.

8.3.3 View

The View menu is a means of granting access to the end-user to find files on the hard drive without actually leaving the application. The menus enables files on had drives to get loaded into the applications designated space. So the the menu has become one of the standard menus that must be included in all applications that involves access to materials on the Local Hard Drive.

As is with the Edit menu, the View menu with the submenu, LineNos, follows the same structure as earlier developed for the File menu and is as given in the follow code:

```
self.vwViewmenu = ToolStripMenuItem("View")
self.vwLineNosmenu = ToolStripMenuItem("Lines No.")
self.vwLineNosmenu.Click += EventHandler(self.SetLineNumbers)
self.vwViewmenu.DropDownItems.Add(self.vwLineNosmenu)
self.mstrip.MdiWindowListItem = self.vwViewmenu
self.mstrip.items.Add(self.vwViewmenu)
```

Again as with the Edit menu codes, this set of codes is inserted, relative to the File menu codes; in the section marked for "optional menus codes" so that the first three and the last two codelines continue to apply to these codes as well.

8.3.4 Tools

The File menu is a means of granting access to the end-user to find files on the hard drive without actually leaving the application. The menus enables files on had drives to get loaded into the applications designated space. So the the menu has become one of the standard menus that must be included in all applications that involves access to materials on the Local Hard Drive.

The essential codes for creating a File menu is as given in the follow code:

8.3.5 Configure

The Configure Menu of main windows when included usually allows access to the ConfigFile through the preferences selection file presented. The menu generally allows the user- or post-installation configuring of the software to the taste of the end-user. Effectively then the option to configure requires that the developer starts off with the inclusion of a ConfigFile, and then also Preference File.

Generally from GUI development stance the ConfigFile development is extratneous, being developed by the information components developer. That then brings up the GU development task to the development of the Preference Files.

Usually a Configure GUI opens up on click of the menu and the image as shown in Fig.8.1 may opens up and let the user make choices of configuring the software;

8.3.6 Help

The Help menu often only has two submenus: UserManual and About. The former usually opens a Pdf document and a GUI develop only needs to point the EventHandler to the location of the PdF book. the latter on the other had opens up a GUI and informs on the legals and Intellectual property matters of the software. Very often the GUI for the legals has the form:

```
def AboutDataDistillery(self, sender, eventargs):
    softwaredescript = "Collation Data Dictillery, ver 1.01; \
    Copyright 2021, Induss SoftWerks. All rights reserved"
    caption = "SoftWerks Collation Data Distillery"
    #msgboxbutton = WinsForms.MessageBoxButton.OK
    #icon = WinsForms.MessageBoxImage.Information
    WinsForms.MessageBox.Show(softwaredescript, caption, 0, 64)
    return True
```

8.4 Toolbar Development

Toolbars as is generally known proffers easy-access to submenus and are productivity enhancement tools, and so coded into just about every Main Window of an application. Toolbars are normally set up with methods of the class objects of the MainWindow as with:

```
def SetupToolBars(self):
    self.tscontner = ToolStripContainer()
    self.tspTop = ToolStripPanel()
    self.tspTop.Dock = WinsForms.DockStyle.Top

    self.ts = ToolStrip()
    self.ts.Location = WinsDraw.Point(0,0)
    self.ts.Size = WinsDraw.Size(self.Width, 25)

    self.tsbButton1 = WinsForms.ToolStripButton()
    self.tsbButton1.Text = str("Button1")
    self.tsbButton1.Tag = str("tsbButton1")
    self.tsb.Size = WinsDraw.Size(int(len("Button1"), 30)
    self.tsb.BorderStyle = FormBorderStyle.FixedDialog
    #self.tsbButton1.Click += self.TextEmailermenu
    self.ts.Items.Add(self.tsbButton1)
    self.tsPanelTop.Join(self.ts)

    self.tsbButton2 = WinsForms.ToolStripButton()
    self.tsbButton2.Text = str("Button2")
    self.tsbButton2.Tag = str("tsbButton2")
    self.tsb.Size = WinsDraw.Size(int(len("‘Button1"’)), 30)
    self.tsb.BorderStyle = FormBorderStyle.FixedDialog
    #self.tsbButton2.Click += self.HTMLEmailermenu
    self.ts.Items.Add(self.tsbButton2)
```

```
        self.tsPanelTop.Join(self.ts)

        self.tsPanelTop.Join(self.ts)
        self.tscontner.TopToolStripPanel.Controls.Add(self.ts)
        self.tscontner.ContentPanel.Controls.Add(self.tsPanelTop)
        self.Controls.Add(self.ts)
        return True
```

Clearly the mounting of the Toolbar as can be surmised is rather static, being mounted with the application startup and before interaction with the end-user. Also, notably the variable ts embodies just one set of toolbar objects that may be associated with a specific class of menu. and other toolbar sets could just as well be created as with this case:

```
    def SetupToolBars(self):
        self.tscontner = ToolStripContainer()
        self.tspTop = ToolStripPanel()
        self.tspTop.Dock = WinsForms.DockStyle.Top

        self.ts1 = ToolStrip()
        self.ts.Location = WinsDraw.Point(0,0)
        self.ts.Size = WinsDraw.Size(self.Width, 25)
        self.tsbButton1 = WinsForms.ToolStripButton()
        self.tsbButton1.Text = str("Button1")
        self.tsbButton1.Tag = str("tsbButton1")
        self.tsb.Size = WinsDraw.Size(int(len("Button1")), 30)
        self.tsb.BorderStyle = FormBorderStyle.FixedDialog
        #self.tsbButton1.Click += self.TextEmailermenu
        self.ts1.Items.Add(self.tsbButton1)
        self.tsPanelTop.Join(self.ts1)

        self.tsPanelTop.Join(self.ts1)
```

```
        self.tscontner.TopToolStripPanel.Controls.Add(self.ts1)
        self.tscontner.ContentPanel.Controls.Add(self.tsPanelTop)
        self.Controls.Add(self.ts1)

        self.ts2 = ToolStrip()
        self.ts2.Location = WinsDraw.Point(
                                 int(len(self.tsbButton1.Text)), 0)
        self.ts2.Size = WinsDraw.Size(self.Width/2, 25)
        self.tsbButton1 = WinsForms.ToolStripButton()
        self.tsbButton1.Text = str("Button1")
        self.tsbButton1.Tag = str("tsbButton1")
        self.tsb.Size = WinsDraw.Size(int(len("Button1")), 30)
        self.tsb.BorderStyle = FormBorderStyle.FixedDialog
        #self.tsbButton1.Click += self.TextEmailermenu
        self.ts2.Items.Add(self.tsbButton1)
        self.tsPanelTop.Join(self.ts2)

        self.tsPanelTop.Join(self.ts2)
        self.tscontner.TopToolStripPanel.Controls.Add(self.ts2)
        self.tscontner.ContentPanel.Controls.Add(self.tsPanelTop)
        self.Controls.Add(self.ts2)
        return True
```

8.5 StatusBar Development

An information GUI object that is often overlooked is the StatusBar. The object provides a means of informing the end-user of the status of the performance of the application as well as of specific functions that are being performed while the application is in use.

Accordingly the coding for a statusbar is significant in the development of any GUI, and the standard coding is given here, in the form in which its usually embedded in the Application MainWindow as a Nested Class

```
class ExampleStstusBar(Form):
    def __ini__(self):
        Form.__init__(self)
        self.Name = "Form1"
        self.ClientSize = WinsDraw.Size(292, 273)
        self.statusStrip1 = None
        self.toolStripStatusLabel1 = None
        self.SetUpStatusBar()

    def SetUpStatusBar(self):
        self.statStrip = WinsForms.StatusStrip()
        self.statStrip.SuspendLayout()
        self.SuspendLayout()
        #// statStrip
        self.statStrip.Location = WinsDraw.Point(0, int(self.Height - 30))
        self.statStrip.Name = str("statusStrip1")
        self.statStrip.Text = str("statusStrip1")
        self.statStrip.Size = System.Drawing.Size(self.Width, 25)
        self.statStrip.BackColor = Color.Navy
        self.statStrip.TabIndex = 0
        #//
        self.tsStatusLabel = WinsForms.ToolStripStatusLabel()
        self.tsStatusLabel.BorderStyle = WinsForms.Border3DStyle.Raised
        self.tsStatusLabel.IsLink = true
        self.tsStatusLabel.Name = str("tsStatusLabel")
        self.tsStatusLabel.Size = WinsDraw.Size(246, 20)
        self.tsStatusLabel.Spring = true
        self.tsStatusLabel.Text = str("tsStatusLabel")
```

```
        self.toolStripStatusLabel1.Alignment = ToolStripItemAlignment.L

        self.Controls.Add(statusStrip1)

        self.statStrip.ResumeLayout(False)
        self.statStrip.PerformLayout()
        self.ResumeLayout(False)
        self.PerformLayout()
        self.Show()
        return True
```

this code is then usually invoked in the MainWindow Initialization function like so:

```
self.statbar = self.ExampleStatusBar(self)
```

and the Statusbar then affixes itself to the MainWindow. the developer then writes to the StatusBar at the various sections where the developer opts to inform the user of the status of the operations

8.6 The Application Main Window Code

The Main Window of an application effectively is the generic Document Interface as shown in Fig. 8.1 augmented with additional menus, toolbars, and function-specific features.

full of packages which turns out to be very helpful to developers as every task does not have to performed from scratch. so then the start of every python development coding is the selection of the packages that needs to be imported into the program for use in the development.

```
import guipymode
from guipymod import *
import huwinsforms
```

```
from guiwinsforms import *
import formbase.guilayout
from formbase.guilayout import SetUpGUIForm

class pyNetBasicProgram(Form):
    def __init__(self):
        self.SetUpGUIForm(self)
        self.SetUpMenus()
        self.SetupToolBars()
        self.Controls.Add(self.ts)
        self.Controls.Add(self.mstrip)
        self.SetUpStatusBar(self):
        self.setdatapanel()
        self.SelfShow((self)

    def SetUpMenus(self):
        self.tscontner = ToolStripContainer()
        self.tspTop = ToolStripPanel()
        self.tspTop.Dock = WinsForms.DockStyle.Top
        self.filemenu = ToolStripMenuItem("File")
        self.fileOpenmenu = ToolStripMenuItem("Open")
        self.fileOpenmenu.Click += self.FileOpenClick
        self.filemenu.DropDownItems.Add(self.fileOpenmenu)
        self.mstrip.MdiWindowListItem = self.filemenu
        self.mstrip.items.Add(filemenu)

        self.edEditmenu = ToolStripMenuItem("Edit")
        self.editCopymenu = ToolStripMenuItem("Copy")
        self.editCopymenu.Click += EventHandler(self.EditCopy)
        self.edEditmenu.DropDownItems.Add(self.editCopymenu)
        self.mstrip.MdiWindowListItem = self.edEditmenu
```

```
        self.mstrip.items.Add(edEditmenu)

        self.vwViewmenu = ToolStripMenuItem("View")
        self.vwLineNosmenu = ToolStripMenuItem("Lines No.")
        self.vwLineNosmenu.Click += EventHandler(
                                            self.SetLineNumbers)
        self.vwViewmenu.DropDownItems.Add(self.vwLineNosmenu)
        self.mstrip.MdiWindowListItem = self.vwViewmenu
        self.mstrip.items.Add(vwViewmenu)

        \emph{---optional more menus---}
        self.MainMenuStrip = self.mstrip


    def SetupToolBars(self):
        self.tscontner = ToolStripContainer()
        self.tspTop = ToolStripPanel()
        self.tspTop.Dock = WinsForms.DockStyle.Top

        self.ts = ToolStrip()
        self.ts.Location = WinsDraw.Point(0,0)
        self.ts.Size = WinsDraw.Size(self.Width, 25)

        self.tsbButton1 = WinsForms.ToolStripButton()
        self.tsbButton1.Text = str("Button1")
        self.tsbButton1.Tag = str("tsbButton1")
        #self.tsbButton1.Click += self.TextEmailermenu
        self.ts.Items.Add(self.tsbButton1)
        self.tsPanelTop.Join(self.ts)

        self.tsbButton2 = WinsForms.ToolStripButton()
```

```
        self.tsbButton2.Text = str("Button2")
        self.tsbButton2.Tag = str("tsbButton1")
        #self.tsbButton2.Click += self.HTMLEmailermenu
        self.ts.Items.Add(self.tsbButton2)
        self.tsPanelTop.Join(self.ts)

        self.tsPanelTop.Join(self.ts)
        self.tscontner.TopToolStripPanel.Controls.Add(self.ts)
        self.tscontner.ContentPanel.Controls.Add(self.tsPanelTop)
        return True

    def SetUpStatusBar(self):
        self.statStrip = WinsForms.StatusStrip()
        self.statStrip.SuspendLayout()
        self.SuspendLayout()
        self.statStrip.Location = WinsDraw.Point(0,
                                     int(self.Height - 30))
        self.statStrip.Name = str("statusStrip1")
        self.statStrip.Text = str("statusStrip1")
        self.statStrip.Size = System.Drawing.Size(
                                     self.Width, 25)
        self.statStrip.BackColor = Color.Navy
        self.statStrip.TabIndex = 0

        self.tsStatusLabel = WinsForms.ToolStripStatusLabel()
        self.tsStatusLabel.BorderStyle = WinsForms.
                                        Border3DStyle.Raised
        self.tsStatusLabel.IsLink = true
        self.tsStatusLabel.Name = str("tsStatusLabel")
        self.tsStatusLabel.Size = WinsDraw.Size(246, 20)
        self.tsStatusLabel.Spring = true
```

```
        self.tsStatusLabel.Text = str("tsStatusLabel")
        self.tsStatusLabel.Alignment =
                              ToolStripItemAlignment.Left

        self.Controls.Add(statStrip)

        self.statStrip.ResumeLayout(False)
        self.statStrip.PerformLayout()
        self.ResumeLayout(False)
        self.PerformLayout()
        return True

    def setdatapanel(self):
        self.datapanel = WinsForms.Panel()
        self.datapanel.BackColor = Color.Navy
        self.datapanel.ForeColor = Color.White
        self.datapanel.Width = self.Width
        self.datapanel.Height = self.Height
        self.datapanel.BorderStyle = BorderStyle.FixedSingle

    def SelfShow((self):
        self.Show()
        return True

    def __del__(self):
        pass
```

The above Application Main Window code has been assembled from the codes presented in the earlier Chapters and is a starting design code for the use by owner-purchaser of the book, ans may be used for developing whatever application for which there is need. however, the adopter of the code may still perform quite an extensive development coding before it should be functional for the needs

8.7 Context Menus Programming

GUI objects have different properties as have been expansively addressed in Chapter 4, 5 and 6 and choice of which are not always obvious. A Context Menu which is a dialog object for learning of and selecting specific properties of a given object enables effective use of the application and so is often included with the software.

However, although by most of the GUI objects have default Context Menus, not all objects –RichTextBox and Picture objects for instance; have associated default Context Menu; and so these objects the developer must explicitly develop Context Menus. For a GUI object to have an association with Context Menu, additional steps beyond the GUI object design specifications as per Chapter ?? have to be taken to ensure that the code executes.

First of, in the GUI Form object design function bearing the ichTextBox Object, say rtbExample, are added the following codes:

```
def __init__(self):
    -----form design codes---
    self.rtbCM = None
    self.InitializeCM()
    self.Show()
    return True
def InitializeCM(self):
    self.cmObj = self.parent.RTBContextMenu(self.parent)
    self.rtbCM = self.cmObj.sbicContextMenuStrip
    return True
```

then in the function bearing the codes for the design specification of the RichTextBox object is added the code assigning the specification to ContextMenu property of the RichTextBox object:

```
def InstanceRichTextBox(self):
    ------design specifiction codes------
    self.rtbExample.ContextMenuStrip = self.rtbCM
    self.Controls.Add(self.rtbExample)
    return True
```

The Context Menu being associated with the RichTextBox object is itself then implemented as a separate class within GUI Form object with these this codes:

```
class RTBContextMenu(Form):
    def __init__(self, parent):
        Form.__init__(self)
        self.parent = parent
        self.rtbContextMenuStrip = None;
        self.CreateContextMenu()

    def CreateContextMenu(self):
        #Create a new ContextMenuStrip control.
        self.rtbContextMenuStrip = WinsForms.ContextMenuStrip();
        self.rtbContextMenuStrip.BackColor = Color.LightGray
        self.rtbContextMenuStrip.ForeColor = Color.Black

        self.rtbContextMenuStrip.Opening += self.cmsShowMenu;

        self.ContextMenuStrip = self.rtbContextMenuStrip;
        return True

    def cmsShowMenu(self, sender, CancelEventArgs):
        #// Acquire references to the owning control and item.
        Contrlc = self.rtbContextMenuStrip.SourceControl #as Control;

        #// Clear the ContextMenuStrip control's Items collection.
```

```
        self.sbicContextMenuStrip.Items.Clear();

        self.sbicContextMenuStrip.Items.Add("&Copy");
        self.sbicContextMenuStrip.Items.Add("&Paste");
        self.sbicContextMenuStrip.Items.Add("&Delete");
        self.sbicContextMenuStrip.Items.Add("-");
        self.sbicContextMenuStrip.Items.Add("&Clear");
        self.sbicContextMenuStrip.Items.Add("&Close");

        #// Set Cancel to false
        CancelEventArgs.Cancel = False;
        return True

    def __del__(self):
        self.Dispose()
```

The code of the Deconstructor, *__del__* is necessary because the opening the ContextMenu is dynamic and so creates the need to also close dynamically thereby necessitating a recovery of the resources which is usually effected with the "Dispose()" function.

Now then placing the mouse pointer on the RichTextBox and right-clicking brings up the Context Menu as shown in Fig. 8.4 and from which a user chooses the menu of interest at the specific time of clicking Each Event of the contextmenu has to be associated with the corresponding Event of the Main Window, and the developer must necessarily implement such.

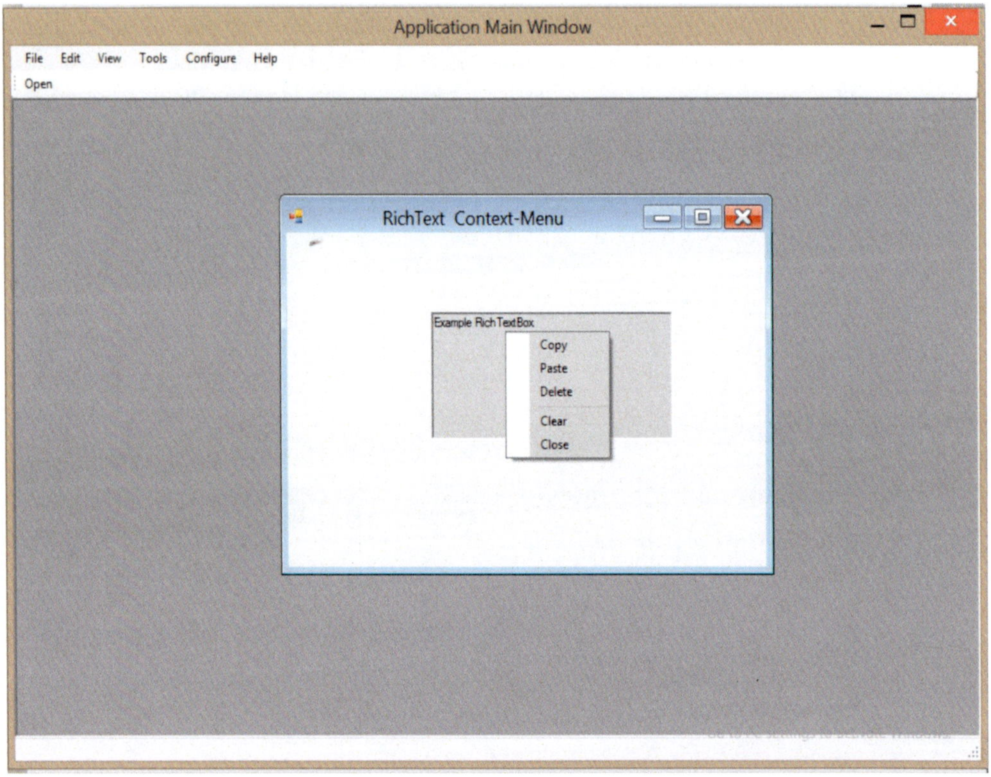

Figure 8.4: RichTextBox ContextMenu Example pythonNet GUI

Chapter 9

Applications PythonNet GUI-Retrofit

9.1 GUI Content Structure Design

By and large, developing pythonNet GUI is the judicious creative integration of the Forms objects that have been described in chapter 5 and 6 in the context of an Application Main Wincow as presented in chapter 8, followed with associating the objects with the related codes on an interactive manner.

Yet in so doing there is still the need for grouping the codes in specific form, reflective of the often adopted form of classes design as is shown in the design:

```
class ClassName:
{
    Members:
        properties and assignment list

    Methods
        methods collection

    Helper Functions
```

```
        special helper Functions

    Bridge Functions
        class-pair relating functions
}
```

Obviously the design of the class separates out the members, and the methods, and then further specializes other members into Helper Functions and Bridge Functions. Naturally then this format of class characteristic grouping ought be opted for, and accordingly implemented for the design of pythonNet GUI class.

9.2 GUI Content Logical Design

Yet the design of a GUI class in accordance with the general class design form comes with a caveat. However noting that the members of a class design are equivalent to properties in GUI class design and custom methods substitute the class behaviour methods, plus the interaction of the end-user is reflected in Events and then addressed yet again with a set of special methods, the EventHandlers, then suffice that the pythonNet GUI class has the design of the type shown now;

```
class GUI-Object
{
    properties;

    Custom Methods:

    EventHandler Function

    Helper Functions
        special helper Functions

    Bridge Functions
```

```
        class-pair relating functions


    Friend Functions:
        groupog Friend functions
}
```

So then the above is the suggested design for a GUI Dialog class into which the codes are advisedly implemented.

The PythonNet GUI Window-class design that is then embedded between the Menu Designs and the Destructor function has the structure like so:

```
class GUI-DialogName:

    Constructor
    default Constructor()
    properties and assignment list
    initialization Constructors()

    Dialog GUI-objects Designs:

    objects design specifications
    -- object properties

    objects EventHandlers
    --- objects EventCancelHandlers

    object mount methods
    --object design specification implementation

    class methods implementation:
    class Custom methods:
    methods collection
```

```
CancelEventHandlers:
    set of Event handlers

Helper Functions
    special helper Functions

Bridge Functions
    class-pair relating functions

default Destructor()
```

Then, of course, the coding for Friend Functions and Bridge Functions are not readily elicited in this case of single dialog GUI designs, where there is absence of class-class interactions or exchange of data.

Particularly noteworthy though are the Object Mount Functions. these functions are not explicitly coded by the developer but rather are built-in method functions that the developer must invoke for the implementation of the GUI objects. these functions are generally of the structure:

```
formobject.Controls.Add(formobject)
```

These formobject have specific relationships for the Add method to apply.

The approach by which the individual pieces of GUI objects are integrated and layout to produce a software GUI for an application can be captured and reflected in a template structure:

- Create SubMenu Folders
- Write one main file for each SubMenu
- Submenu File Coding: Always start with panel, then objects, and finalize layout design with mount layout
- Interaction/Interactivity Design, EventHandlers for the objects

For effect and insight, a simple case of Big Data use of the utility of MongoDB is presented

9.3 Retrofitting GUI

The development through the previous chapters are explicitly implemented here to the develop the GUI for existing Console applications. In particular, the retrofit is performed on Client-server Architecture Systems.

In keeping with the instructions, the approach essential consists of grouping the Client into functions-sets consistent with the standard C Programming constructs consisting of the three key functions: DataInput function, Application Function, Display Function, such as

```
def Display(results):
    print(r"'The result is: %12s"'%(results))
    return True

def DataInput():
    put("Enter Some Data and press Enter:" )
    return True

def PerformOps(dataentery):
    return True
```

After this basic construct is met then all other functions that perform some operations on the input data whether directly or indirectly are grouped into sets. Then all functions that directly invoke the DataInput and Display Functions are also grouped into EventHandlers and the rest grouped as Custom Methods for invocation in the EventHandler functions.

Finally pythonNet GUI is developed for utilizing the DataInput and Display functions as is appropriate for the data needs, and for executing the EventHandlers.

9.4 Socket Client-Server Client-GUI

The common cases of applications are the Client-Server and Distributed Computing Systems that remote the operations so that much larger system collection can be used for the operations.

Although several different InterProcess Communication, IPC,tools are sometimes used for this purpose, Socket Client-Server application is the most rudimentary option often used. So it is reasonable to start with the development of a GUI for for the client side of the application; besides such is particularly instructive in the development of GUI for all such applications irrespective of the actual architecture of the Client-Server or Distributed Applications.

The Socket Client-Server is particularly apt for developments towards concept validation of server development. A particularly useful form is in the development of LAN based RPC where the computing functions are executed on the server byt actually invoked on the client. although this approach may not quite suited for production operations it serves well for the development of collection of computational functions that could be later deployed on a production server.

In that regard therefore a GUI for such Client Server system is just as useful for later integration into defining the design of the Clients for Production systems. A simple but guiding code implementation of development such GUI may not to too far from the code for an addition Calculator:

```
\#//Global call interface function
from __future__ import division
import sys \#for exit
import math
import time
import numpy
from numpy import *
import scipy
from scipy.integrate import simps
import sympy
```

```
import matplotlib.pyplot as plt
import ast
import socket #for sockets
import pywintypes
import win32ui
import win32con \#constants
import ctypes
import os as os

class TCSocketClient:
    def __init__(self):
        self.host = None
        self.port = None
        self.socket = None
        selffnum = 0
        selfsnum = 0
        self.InitializeSocket()

    def InitializeSocket(self):
        try:
            #create an AF_INET, STREAM socket (TCP)
            s = socket.socket(socket.AF_INET, socket.SOCK_STREAM)
         except socket.herror as msg:
            print(r"Failed to create socket. Error code: {:34s},
                  Error message : {:34s}".format(msg[0], msg[1]))
            sys.exit()
        print(r" {:34s}".format('Socket Created'))

        host = 'localhost'
        port = 8888
        try:
```

```
            remote_ip = socket.gethostbyname( host )
        except socket.gaierror:
            \#could not resolve
            print(r" {:34s}".format('Hostname could
                not be resolved. Exiting'))
            sys.exit()
        print(r"Ip address of {:34s} is {:34s}".format(
                                host,remote_ip))

        #Connect to remote server
        s.connect((remote_ip , port))
        print(r"Socket Connected to {:34s} on ip {:34s}".format(
                                           host,remote_ip))
        self.socket = s
        return True

    def ServerRequest(self, message):
        s = self.socket
        \#Send some data to remote server
        try:
            \#Set the whole string
            s.sendall(message.encode("utf-8"))
            \#s.write(message.encode("utf-8"))
            \# alternatively simply enter the opsData
        except socket.herror:
            \#Send failed
            print(r" {:34s}".format('Send failed'))
            sys.exit()
        print(r" {:34s}".format('Data send: Successful'))
        return True
```

```
    def ServerResponse(self):
        s = self.socket
        reply = s.read(4096)
        print(r" {:34s}".format(reply.decode("utf-8")))
        return reply

    def __del__(self):
        s.close()

    def DataInput():
        datalst = []
        fnum = input("Enter first number and press enter: ")
        snum = input("Enter second number and press enter: ")
        datalst .append(fnum)
        datalst.append(snum)
        return datalst

def main():
    #insert socketclient here
    dataset = []
    dataset = DataInput()
    print(r"Processing the Server Request ...
                    please wait...")
    clsocket = TCSocketClient() #then display results
    msgstr = str(dataset[0]) + str(",") + str(dataset[1])
    res = clsocket.ServerRequest(msgstr)
    data = clsocket.ServerResponse()
    print(r"Server Response to Server Request:\%34s"\%(data))
    return True
```

The object then is to embedd this datasets into a pythonet GUI . At this point of the development, the decision about the architecture needs be made and the developer may so choose as preferred. For simplicity though and reuse of earlier concepts the container design of section ?? is chosen: One section allows the entry and sending off of the data to the server and the other section displays the results.

Effectively the GUI has the appearance as shown in Fig. ?? where the Windows Forms are created without the FixedBorder option and as such the Form is

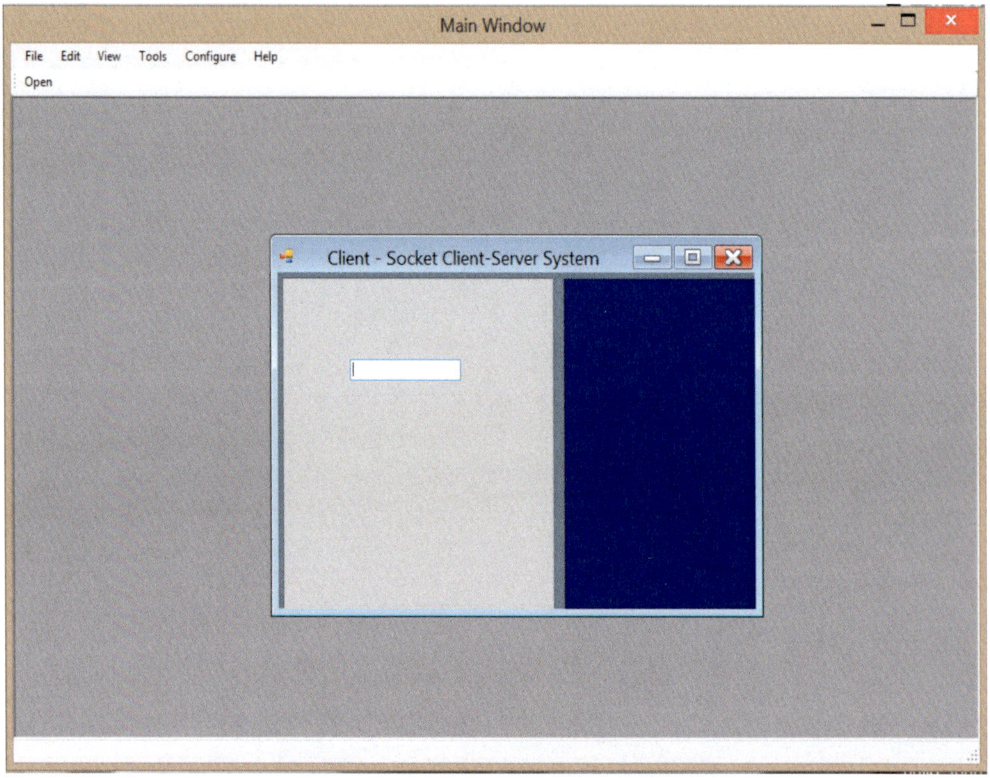

Figure 9.1: A Basic pythonNet GUI

flat in the underlying panels. For embedding the code in a pythonet GUI, first off, the gRPC client code is saved, say,in a folder named "socketclient" and the file itself named "socketclient.py". Next, the main program for executing the Client is copied into the GUI class and renamed as a function, say, SendData(self). All and all, the finished GUI code is as:

```
import headers
from headers import *
import formbase.guilayout
from formbase.guilayout import SetUpGUIForm

import socketclient.socketclient
from socketclient.socketclient import TCSocketClient
class CalculatorClient(Form):
    def __init__(self):
        self.SetUpGUIForm(self)
        self.InstanceLabel()
        self.GUIFormShow()

    def InstanceGUIObjects(self):
        self.txtExample = WinsForms.TextBox()
        self.txtExample.Text = str("")
        self.txtExample.Location = WinsDraw.Point(70, 75)
        self.txtExample.Height = 30
        self.txtExample.Width = 115
        self.txtExample.BackColor = Color.LightGray
        self.txtExample.ForeColor = Color.Navy
        #self.txtExample.HorizontalAlignment =
                             WinsForms.HorizontalAlignment.Center
        self.lPanel.Controls.Add(self.txtExample)

        self.txtExample = WinsForms.TextBox()
        self.txtExample.Text = str("")
        self.txtExample.Location = WinsDraw.Point(70, 100)
        self.txtExample.Height = 30
        self.txtExample.Width = 115
        self.txtExample.BackColor = Color.LightGray
```

```
        self.txtExample.ForeColor = Color.Navy
        #self.txtExample.HorizontalAlignment =
                            WinsForms.HorizontalAlignment.Center
        self.lPanel.Controls.Add(self.txtExample)
        return True

    def SendData(self):
        #insert socketclient here
        msgstr = self.fnum + str(",") + self.snum
        clsocket = TCSocketClient() #then display results
        res = clsocket.ServerRequest(msgstr)
        statusmsg = str("Processing the Server Request ...
                                please wait...")
        self.container.disDlg.Write(statusmsg)
        data = clsocket.ServerResponse()
        self.Receive(data)
        return True

    def Onsend(self, sender, eventargs):
        self.fnum = self.txtFirstNumber.Text
        self.snum = self.txtSecondNumber.Text
        self.SendData()
        return True

    def Receive(self, results):
        retval = results
        responsemsg = str("Server Response to the Server
                                    Request :") + str(retval)
        self.container.disDlg.Write(responsemsg)
        return True
```

```
    def __del__(self):
        pass
```

The GUI of Fig. ?? in effect bears the codes of the design specifications as observed, however, a corresponding server application will have to developed for full use of he design but the objective here is in essence a Thi Client.

9.5 gRPC Client-Server GUI-Client

A rather interesting context of use of pythonnet GUI is the scientific computation of sorts. There are myriad applications for which such GUIs are useful. Often the computational algorithms are installed in remote server that are quite power but available for the user to access by simply sending data into the system and then receiving the results of the computation in a round trip. RPC has been the method often chosen for such Client Server Applications.

While varied approaches have been used in the implementation of the RPC Client-server systems, the one of presentation here is the gRPC[8] which is the RPC model developed by and in Google and made available to the general public. Though complicated uses of this system have been developed however, being an illustrative development the presentation here is simply for the addition of two number, the foundation for the development of a calculator for performing complex computation remotely.

A typical client class of the gRPC saved as *ghello.py* for computing two number has the construct as

```
\#ghelloclient.py
from __future__ import division
import sys \#for exit
import socket \#for sockets
import math
import time
```

```
import grpc
import greethello_pb2
import greethello_pb2_grpc
from grpc import Call as call

class Hello:
    def __init__(self, name, id):
        self.name = name
        self.id = id
        self.message = None
        self.channel = None
        self.stub = None
        self.greetings = None
        self.chnfob = None
        self.response = str("")
        self.InitializeHello()

    class Greetings:
        def __init__(self, parent):
            self.name = parent.name
            self.id = parent.id

        def __del__(self):
            pass

    def InitializeHello(self):
        grpcVersion = grpc.__version__
        serveraddr = 'localhost:50051' #set this by config file
        \#---------
        \# creates a client object: channel defines client to server
        channel = grpc.insecure_channel(serveraddr, None, None)
```

```
        channel.__enter__()
        self.channel = channel
        self.chnfob = grpc.channel_ready_future(channel)
        \#--------
        self.stub = greethello_pb2_grpc.HelloStub(self.channel)
        lcred = grpc.local_channel_credentials(grpc.LocalConnectionType.LC
        \#key, value = call.trailing_metadata()
        \#print(r"key %34s value %34s"%(key, value))
        self.greetings = self.Greetings(self)
        return True

    def HelloRequest(self):
        \#---grpc.ChannelConnectivity(self.chnfob)
        \#---print(r"connvalue %34s"%(self.chnfob))
        \#if self.chnfob == grpc.ChannelConnectivity.READY:
        """ insert a metadata on wait_for_ready so the chanel gets ready ,
        \#-------
        \# create a valid request message
        \#assign to parameter to make work
        message = greethello_pb2.Greetings(name=self.greetings.name, id=se
        reqmsg = greethello_pb2.HelloRequest(name=message.name)
        stub = self.stub
        response = stub.SayHello(request=reqmsg)
        self.response = response
        try:
            pass
        except grpc.RpcError as rpc_error_call:  \# object is indeed a grp
            my_code = rpc_error_call.code()
            my_details = rpc_error_call.details()
        return True
```

```
    def HelloReply(self):
        response = self.response
        return response

    def __del__(self):
        \#self.channel.__exit__()
        \# requires 3 positional arguments
        pass

def DataInput():
    datalst = []
    fnum = str("You").encode("utf-8")
    snum = input("Enter the first number and press enter:  ")
    datalst .append(fnum)
    datalst.append(int(snum))
    return datalst

def main():
    dataset = DataInput()
    grob = Hello(dataset[0], dataset[1])
    grob.HelloRequest()
    grob.HelloReply()
    return True
```

The object then is to embedd this datasets into a pythonet GUI . At this point of the development, the decision about the architecture needs be made and the developer may so choose as preferred. For simplicity though and reuse of earlier concepts the the container design of section ?? is chosen: One section allows the entry and sending off of the data to the server and the other section display the results.

Effectively the GUI has the appearance as shown in Fig. ?? where the Windows Forms are created without the FixedBorder option and as such the Form is

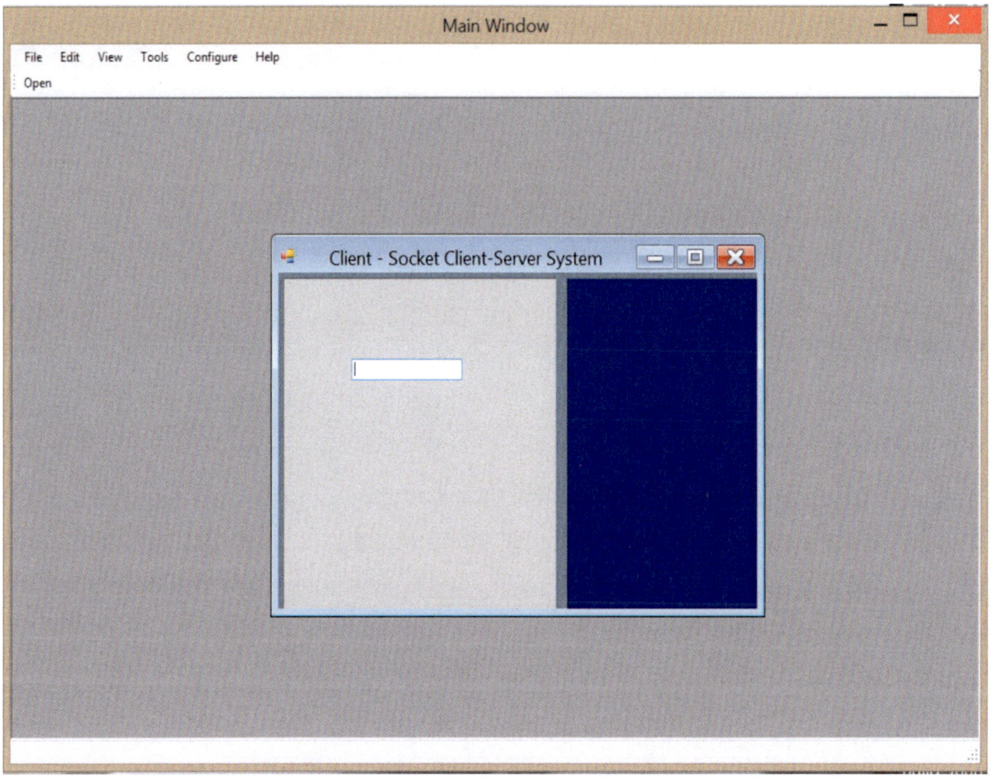

Figure 9.2: A Basic pythonNet GUI

flat in the underlying panels. For embedding the code in a pythonet GUI, first off, the gRPC client code is saved, say,in a folder named "ghello" and the file itself named "ghello.py". Next, the main program for executing the Client is copied into the GUI class and renamed as a function, say, SendData(self). All and all, the finished GUI code is as:

```
import headers
from headers import *
import formbase.guilayout
from formbase.guilayout import SetUpGUIForm
```

```
import ghello.ghello
from ghello.ghello import Hello

class GreeterClient(Form):
    def __init__(self):
        self.SetUpGUIForm(self)
        self.InstanceLabel()
        self.GUIFormShow()

    def SendData(self):
        grob = Hello(self.name, self.id)
        grob.HelloRequest()
        statusmsg = str("Processing the Server Request ...please wait..
        self.container.disDlg.Write(statusmsg)
        retvals = grob.HelloReply()
        self.receive(retvals)
        return True

    def Onsend(self, sender, eventargs):
        self.name = int(self.lblFirstNumber.Text)
        self.id = int(self.txtIDData.Text)
        self.SendData()
        return True

    def Receive(self, results):
        retval = results
        responsemsg = str("Server Response of the Server Request :") +
        self.container.disDlg.Write(responsemsg)
        return True

    def InitializeHello(self):
```

```
            self.lblFirstName.Text = str("You")
            return True

        def __del__(self):
            pass
```

Upon the entry of the data and followed by the click of the button Compute, the data flows out to the server and after a short while the results display on the right Form.

Appendix A - Properties

AccessibilityObject
Gets the AccessibleObject assigned to the control.
(Inherited from Control)

AccessibleDefaultActionDescription
Gets or sets the default action description of the control for use by accessibility client applications.
(Inherited from Control)

AccessibleDescription
Gets or sets the description of the control used by accessibility client applications.
(Inherited from Control)

AccessibleName
Gets or sets the name of the control used by accessibility client applications.
(Inherited from Control)

AccessibleRole
Gets or sets the accessible role of the control.
(Inherited from Control)

AllowDrop
Gets or sets a value indicating whether the control can accept data that the user drags onto it.

Anchor
Gets or sets the edges of the container to which a control is bound and determines how a control is resized with its parent.

(Inherited from Control)

AutoScrollOffset
Gets or sets where this control is scrolled to in ScrollControlIntoView(Control).
(Inherited from Control)

AutoSize
This property is not relevant for this class.
(Inherited from Control)

BackColor
Gets or sets the background color for the control.
(Inherited from Control)

BackgroundImage
Gets or sets the background image for the ProgressBar control.

BackgroundImageLayout
Gets or sets the layout of the background image of the progress bar.

BindingContext
Gets or sets the BindingContext for the control.
(Inherited from Control)

Bottom
Gets the distance, in pixels, between the bottom edge of the control and the top edge of its container's client area.
(Inherited from Control)

Bounds
Gets or sets the size and location of the control including its nonclient elements, in pixels, relative to the parent control.
(Inherited from Control)

CanEnableIme
Gets a value indicating whether the ImeMode property can be set to an active value, to enable IME support.
(Inherited from Control)

CanFocus
Gets a value indicating whether the control can receive focus.
(Inherited from Control)

CanRaiseEvents
Determines if events can be raised on the control.
(Inherited from Control)

CanSelect
Gets a value indicating whether the control can be selected.
(Inherited from Control)

Capture
Gets or sets a value indicating whether the control has captured the mouse.
(Inherited from Control)

CausesValidation
Gets or sets a value indicating whether the control, when it receives focus, causes validation to be performed on any controls that require validation.

ClientRectangle
Gets the rectangle that represents the client area of the control.
(Inherited from Control)

ClientSize
Gets or sets the height and width of the client area of the control.
(Inherited from Control)

CompanyName
Gets the name of the company or creator of the application containing the control.
(Inherited from Control)

Container
Gets the IContainer that contains the Component.
(Inherited from Component)

ContainsFocus

Gets a value indicating whether the control, or one of its child controls, currently has the input focus.
(Inherited from Control)

ContextMenu
Gets or sets the shortcut menu associated with the control.
(Inherited from Control)

ContextMenuStrip
Gets or sets the ContextMenuStrip associated with this control.
(Inherited from Control)

Controls
Gets the collection of controls contained within the control. (Inherited from Control)

Created
Gets a value indicating whether the control has been created.
(Inherited from Control)

CreateParams
Overrides CreateParams.

Cursor
Gets or sets the cursor that is displayed when the mouse pointer is over the control.
(Inherited from Control)

DataBindings
Gets the data bindings for the control.
(Inherited from Control)

DefaultCursor
Gets or sets the default cursor for the control.
(Inherited from Control)

DefaultImeMode
Gets the default Input Method Editor (IME) mode supported by the control.

DefaultMargin
Gets the space, in pixels, that is specified by default between controls.
(Inherited from Control)

DefaultMaximumSize
Gets the length and height, in pixels, that is specified as the default maximum size of a control.
(Inherited from Control)

DefaultMinimumSize
Gets the length and height, in pixels, that is specified as the default minimum size of a control.
(Inherited from Control)

DefaultPadding
Gets the internal spacing, in pixels, of the contents of a control.
(Inherited from Control)

DefaultSize
Gets the default size of the control.

DesignMode
Gets a value that indicates whether the Component is currently in design mode.
(Inherited from Component)
DeviceDpi
Gets the DPI value for the display device where the control is currently being displayed.
(Inherited from Control)

DisplayRectangle
Gets the rectangle that represents the display area of the control.
(Inherited from Control)

Disposing
Gets a value indicating whether the base Control class is in the process of disposing.
(Inherited from Control)

Dock
Gets or sets which control borders are docked to its parent control and determines how a control is resized with its parent.
(Inherited from Control)

DoubleBuffered
Gets or sets a value indicating whether the control should redraw its surface using a secondary buffer.

Enabled
Gets or sets a value indicating whether the control can respond to user interaction.
(Inherited from Control)

Events
Gets the list of event handlers that are attached to this Component.
(Inherited from Component)

Focused
Gets a value indicating whether the control has input focus.
(Inherited from Control)

Font
Gets or sets the font of text in the ProgressBar.

FontHeight
Gets or sets the height of the font of the control.
(Inherited from Control)

ForeColor
Gets or sets the foreground color of the control.
(Inherited from Control)

Handle
Gets the window handle that the control is bound to.
(Inherited from Control)

HasChildren
Gets a value indicating whether the control contains one or more child controls.
(Inherited from Control)

Height
Gets or sets the height of the control.
(Inherited from Control)

ImeMode
Gets or sets the input method editor (IME) for the ProgressBar.

ImeModeBase
Gets or sets the IME mode of a control.
(Inherited from Control)

InvokeRequired
Gets a value indicating whether the caller must call an invoke method when making method calls to the control because the caller is on a different thread than the one the control was created on.
(Inherited from Control)

IsAccessible
Gets or sets a value indicating whether the control is visible to accessibility applications.
(Inherited from Control)

IsDisposed
Gets a value indicating whether the control has been disposed of.
(Inherited from Control)

IsHandleCreated
Gets a value indicating whether the control has a handle associated with it. (Inherited from Control)

IsMirrored
Gets a value indicating whether the control is mirrored.
(Inherited from Control)

LayoutEngine
Gets a cached instance of the control's layout engine.
(Inherited from Control)

Left
Gets or sets the distance, in pixels, between the left edge of the control and the left edge of its container's client area.
(Inherited from Control)

Location
Gets or sets the coordinates of the upper-left corner of the control relative to the upper-left corner of its container.
(Inherited from Control)

Margin
Gets or sets the space between controls.
(Inherited from Control)

MarqueeAnimationSpeed
Gets or sets the time period, in milliseconds, that it takes the progress block to scroll across the progress bar.

Maximum
Gets or sets the maximum value of the range of the control.

MaximumSize
Gets or sets the size that is the upper limit that GetPreferredSize(Size) can specify.
(Inherited from Control)

Minimum
Gets or sets the minimum value of the range of the control.

MinimumSize
Gets or sets the size that is the lower limit that GetPreferredSize(Size) can specify.
(Inherited from Control)

Name
Gets or sets the name of the control.
(Inherited from Control)

Padding
Gets or sets the space between the edges of a ProgressBar control and its contents.

Parent
Gets or sets the parent container of the control.
(Inherited from Control)

PreferredSize
Gets the size of a rectangular area into which the control can fit.
(Inherited from Control)

ProductName
Gets the product name of the assembly containing the control.
(Inherited from Control)

ProductVersion
Gets the version of the assembly containing the control.
(Inherited from Control)

RecreatingHandle
Gets a value indicating whether the control is currently re-creating its handle.
(Inherited from Control)

Region
Gets or sets the window region associated with the control.
(Inherited from Control)

RenderRightToLeft
Obsolete.
This property is now obsolete.
(Inherited from Control)

ResizeRedraw
Gets or sets a value indicating whether the control redraws itself when resized.
(Inherited from Control)

Right
Gets the distance, in pixels, between the right edge of the control and the left

edge of its container's client area.
(Inherited from Control)

RightToLeft
Gets or sets a value indicating whether control's elements are aligned to support locales using right-to-left fonts.
(Inherited from Control)

RightToLeftLayout
Gets or sets a value indicating whether the ProgressBar and any text it contains is displayed from right to left.

ScaleChildren
Gets a value that determines the scaling of child controls.
(Inherited from Control)

ShowFocusCues
Gets a value indicating whether the control should display focus rectangles.
(Inherited from Control)

ShowKeyboardCues
Gets a value indicating whether the user interface is in the appropriate state to show or hide keyboard accelerators.
(Inherited from Control)

Site
Gets or sets the site of the control. (Inherited from Control)
Size
Gets or sets the height and width of the control.
(Inherited from Control)

Step
Gets or sets the amount by which a call to the PerformStep() method increases the current position of the progress bar.

Style
Gets or sets the manner in which progress should be indicated on the progress bar.

TabIndex
Gets or sets the tab order of the control within its container. (Inherited from Control)

TabStop
Overrides TabStop.

Tag
Gets or sets the object that contains data about the control.
(Inherited from Control)

Text
Gets or sets the text associated with this control.

Top
Gets or sets the distance, in pixels, between the top edge of the control and the top edge of its container's client area.
(Inherited from Control)

TopLevelControl
Gets the parent control that is not parented by another Windows Forms control. Typically, this is the outermost Form that the control is contained in.
(Inherited from Control)

UseWaitCursor
Gets or sets a value indicating whether to use the wait cursor for the current control and all child controls.
(Inherited from Control)

Value
Gets or sets the current position of the progress bar.

Visible
Gets or sets a value indicating whether the control and all its child controls are displayed.
(Inherited from Control)

Width

Gets or sets the width of the control.
(Inherited from Control)

WindowTarget
This property is not relevant for this class.
(Inherited from Control)

Appendix B - Methods

AccessibilityNotifyClients(AccessibleEvents, Int32)
Notifies the accessibility client applications of the specified AccessibleEvents for the specified child control.
(Inherited from Control)

AccessibilityNotifyClients(AccessibleEvents, Int32, Int32)
Notifies the accessibility client applications of the specified AccessibleEvents for the specified child control .
(Inherited from Control)

BeginInvoke(Delegate)
Executes the specified delegate asynchronously on the thread that the control's underlying handle was created on.
(Inherited from Control)

BeginInvoke(Delegate, Object[])
Executes the specified delegate asynchronously with the specified arguments, on the thread that the control's underlying handle was created on.
(Inherited from Control)

BringToFront()
Brings the control to the front of the z-order.
(Inherited from Control)

Contains(Control)
Retrieves a value indicating whether the specified control is a child of the control.
(Inherited from Control)

CreateAccessibilityInstance()
Creates a new instance of the accessibility object for the ProgressBar control.

CreateControl()
Forces the creation of the visible control, including the creation of the handle and any visible child controls.
(Inherited from Control)

CreateControlsInstance()
Creates a new instance of the control collection for the control.
(Inherited from Control)

CreateGraphics()
Creates the Graphics for the control.
(Inherited from Control)

CreateHandle()
Creates a handle for the control.

CreateObjRef(Type)
Creates an object that contains all the relevant information required to generate a proxy used to communicate with a remote object.
(Inherited from MarshalByRefObject)

DefWndProc(Message)
Sends the specified message to the default window procedure.
(Inherited from Control)

DestroyHandle()
Destroys the handle associated with the control.
(Inherited from Control)

Dispose()
Releases all resources used by the Component.
(Inherited from Component)

Dispose(Boolean)
Releases the unmanaged resources used by the Control and its child controls and

optionally releases the managed resources.
(Inherited from Control)

DoDragDrop(Object, DragDropEffects)
Begins a drag-and-drop operation.
(Inherited from Control)

DrawToBitmap(Bitmap, Rectangle)
Supports rendering to the specified bitmap.
(Inherited from Control)

EndInvoke(IAsyncResult)
Retrieves the return value of the asynchronous operation represented by the IAsyncResult passed.
(Inherited from Control)

Equals(Object)
Determines whether the specified object is equal to the current object.
(Inherited from Object)

FindForm()
Retrieves the form that the control is on.
(Inherited from Control)

Focus()
Sets input focus to the control.
(Inherited from Control)

GetAccessibilityObjectById(Int32)
Retrieves the specified AccessibleObject.
(Inherited from Control)

GetAutoSizeMode()
Retrieves a value indicating how a control will behave when its AutoSize property is enabled.
(Inherited from Control)

GetChildAtPoint(Point)

Retrieves the child control that is located at the specified coordinates.
(Inherited from Control)

GetChildAtPoint(Point, GetChildAtPointSkip)
Retrieves the child control that is located at the specified coordinates, specifying whether to ignore child controls of a certain type.
(Inherited from Control)

GetContainerControl()
Returns the next ContainerControl up the control's chain of parent controls.
(Inherited from Control)

GetHashCode()
Serves as the default hash function.
(Inherited from Object)

GetLifetimeService()
Obsolete.
Retrieves the current lifetime service object that controls the lifetime policy for this instance.
(Inherited from MarshalByRefObject)

GetNextControl(Control, Boolean)
Retrieves the next control forward or back in the tab order of child controls.
(Inherited from Control)

GetPreferredSize(Size)
Retrieves the size of a rectangular area into which a control can be fitted.
(Inherited from Control)

GetScaledBounds(Rectangle, SizeF, BoundsSpecified)
Retrieves the bounds within which the control is scaled.
(Inherited from Control)

GetService(Type)
Returns an object that represents a service provided by the Component or by its Container.
(Inherited from Component)

GetStyle(ControlStyles)
Retrieves the value of the specified control style bit for the control.
(Inherited from Control)

GetTopLevel()
Determines if the control is a top-level control.
(Inherited from Control)

GetType()
Gets the Type of the current instance.
(Inherited from Object)

Hide()
Conceals the control from the user.
(Inherited from Control)

Increment(Int32)
Advances the current position of the progress bar by the specified amount.

InitializeLifetimeService()
Obsolete.
Obtains a lifetime service object to control the lifetime policy for this instance.
(Inherited from MarshalByRefObject)

InitLayout()
Called after the control has been added to another container.
(Inherited from Control)

Invalidate()
Invalidates the entire surface of the control and causes the control to be redrawn.
(Inherited from Control)

Invalidate(Boolean)
Invalidates a specific region of the control and causes a paint message to be sent to the control. Optionally, invalidates the child controls assigned to the control.
(Inherited from Control)

Invalidate(Rectangle)
Invalidates the specified region of the control (adds it to the control's update region, which is the area that will be repainted at the next paint operation), and causes a paint message to be sent to the control.
(Inherited from Control)

Invalidate(Rectangle, Boolean)
Invalidates the specified region of the control (adds it to the control's update region, which is the area that will be repainted at the next paint operation), and causes a paint message to be sent to the control. Optionally, invalidates the child controls assigned to the control.
(Inherited from Control)

Invalidate(Region)
Invalidates the specified region of the control (adds it to the control's update region, which is the area that will be repainted at the next paint operation), and causes a paint message to be sent to the control.
(Inherited from Control)

Invalidate(Region, Boolean)
Invalidates the specified region of the control (adds it to the control's update region, which is the area that will be repainted at the next paint operation), and causes a paint message to be sent to the control. Optionally, invalidates the child controls assigned to the control.
(Inherited from Control)

Invoke(Delegate)
Executes the specified delegate on the thread that owns the control's underlying window handle.
(Inherited from Control)

Invoke(Delegate, Object[])
Executes the specified delegate, on the thread that owns the control's underlying window handle, with the specified list of arguments.
(Inherited from Control)

InvokeGotFocus(Control, EventArgs)
Raises the GotFocus event for the specified control.

(Inherited from Control)

InvokeLostFocus(Control, EventArgs)
Raises the LostFocus event for the specified control.
(Inherited from Control)

InvokeOnClick(Control, EventArgs)
Raises the Click event for the specified control.
(Inherited from Control)

InvokePaint(Control, PaintEventArgs)
Raises the Paint event for the specified control.
(Inherited from Control)

InvokePaintBackground(Control, PaintEventArgs)
Raises the PaintBackground event for the specified control.
(Inherited from Control)

IsInputChar(Char)
Determines if a character is an input character that the control recognizes.
(Inherited from Control)

IsInputKey(Keys)
Determines whether the specified key is a regular input key or a special key that requires preprocessing.
(Inherited from Control)

LogicalToDeviceUnits(Int32)
Converts a Logical DPI value to its equivalent DeviceUnit DPI value.
(Inherited from Control)

LogicalToDeviceUnits(Size)
Transforms a size from logical to device units by scaling it for the current DPI and rounding down to the nearest integer value for width and height.
(Inherited from Control)

MemberwiseClone()
Creates a shallow copy of the current Object.

(Inherited from Object)

MemberwiseClone(Boolean)
Creates a shallow copy of the current MarshalByRefObject object.
(Inherited from MarshalByRefObject)

NotifyInvalidate(Rectangle)
Raises the Invalidated event with a specified region of the control to invalidate.
(Inherited from Control)

OnAutoSizeChanged(EventArgs)
Raises the AutoSizeChanged event.
(Inherited from Control)

OnBackColorChanged(EventArgs)
Raises the BackColorChanged event.

OnBackgroundImageChanged(EventArgs)
Raises the BackgroundImageChanged event.
(Inherited from Control)
OnBackgroundImageLayoutChanged(EventArgs)
Raises the BackgroundImageLayoutChanged event.
(Inherited from Control)

OnBindingContextChanged(EventArgs)
Raises the BindingContextChanged event.
(Inherited from Control)

OnCausesValidationChanged(EventArgs)
Raises the CausesValidationChanged event.
(Inherited from Control)

OnChangeUICues(UICuesEventArgs)
Raises the ChangeUICues event.
(Inherited from Control)

OnClick(EventArgs)
Raises the Click event.

(Inherited from Control)

OnClientSizeChanged(EventArgs)
Raises the ClientSizeChanged event.
(Inherited from Control)

OnContextMenuChanged(EventArgs)
Raises the ContextMenuChanged event.
(Inherited from Control)

OnContextMenuStripChanged(EventArgs)
Raises the ContextMenuStripChanged event.
(Inherited from Control)

OnControlAdded(ControlEventArgs)
Raises the ControlAdded event.
(Inherited from Control)

OnControlRemoved(ControlEventArgs)
Raises the ControlRemoved event.
(Inherited from Control)

OnCreateControl()
Raises the CreateControl() method.
(Inherited from Control)

OnCursorChanged(EventArgs)
Raises the CursorChanged event.
(Inherited from Control)

OnDockChanged(EventArgs)
Raises the DockChanged event.
(Inherited from Control)

OnDoubleClick(EventArgs)
Raises the DoubleClick event.
(Inherited from Control)

OnDpiChangedAfterParent(EventArgs)
Raises the DpiChangedAfterParent event.
(Inherited from Control)

OnDpiChangedBeforeParent(EventArgs)
Raises the DpiChangedBeforeParent event.
(Inherited from Control)

OnDragDrop(DragEventArgs)
Raises the DragDrop event.
(Inherited from Control)

OnDragEnter(DragEventArgs)
Raises the DragEnter event.
(Inherited from Control)

OnDragLeave(EventArgs)
Raises the DragLeave event.
(Inherited from Control)

OnDragOver(DragEventArgs)
Raises the DragOver event.
(Inherited from Control)

OnEnabledChanged(EventArgs)
Raises the EnabledChanged event.
(Inherited from Control)

OnEnter(EventArgs)
Raises the Enter event.
(Inherited from Control)

OnFontChanged(EventArgs)
Raises the FontChanged event.
(Inherited from Control)

OnForeColorChanged(EventArgs)
Raises the ForeColorChanged event.

OnGiveFeedback(GiveFeedbackEventArgs)
Raises the GiveFeedback event.
(Inherited from Control)

OnGotFocus(EventArgs)
Raises the GotFocus event.
(Inherited from Control)

OnHandleCreated(EventArgs)
Overrides OnHandleCreated(EventArgs).

OnHandleDestroyed(EventArgs)
Raises the HandleDestroyed event.

OnHelpRequested(HelpEventArgs)
Raises the HelpRequested event.
(Inherited from Control)

OnImeModeChanged(EventArgs)
Raises the ImeModeChanged event.
(Inherited from Control)

OnInvalidated(InvalidateEventArgs)
Raises the Invalidated event.
(Inherited from Control)

OnKeyDown(KeyEventArgs)
Raises the KeyDown event.
(Inherited from Control)

OnKeyPress(KeyPressEventArgs)
Raises the KeyPress event.
(Inherited from Control)

OnKeyUp(KeyEventArgs)
Raises the KeyUp event.
(Inherited from Control)

OnLayout(LayoutEventArgs)
Raises the Layout event.
(Inherited from Control)

OnLeave(EventArgs)
Raises the Leave event.
(Inherited from Control)

OnLocationChanged(EventArgs)
Raises the LocationChanged event.
(Inherited from Control)

OnLostFocus(EventArgs)
Raises the LostFocus event.
(Inherited from Control)

OnMarginChanged(EventArgs)
Raises the MarginChanged event.
(Inherited from Control)

OnMouseCaptureChanged(EventArgs)
Raises the MouseCaptureChanged event.
(Inherited from Control)

OnMouseClick(MouseEventArgs)
Raises the MouseClick event.
(Inherited from Control)

OnMouseDoubleClick(MouseEventArgs)
Raises the MouseDoubleClick event.
(Inherited from Control)

OnMouseDown(MouseEventArgs)
Raises the MouseDown event.
(Inherited from Control)

OnMouseEnter(EventArgs)

Raises the MouseEnter event.
(Inherited from Control)

OnMouseHover(EventArgs)
Raises the MouseHover event.
(Inherited from Control)

OnMouseLeave(EventArgs)
Raises the MouseLeave event.
(Inherited from Control)

OnMouseMove(MouseEventArgs)
Raises the MouseMove event.
(Inherited from Control)

OnMouseUp(MouseEventArgs)
Raises the MouseUp event.
(Inherited from Control)

OnMouseWheel(MouseEventArgs)
Raises the MouseWheel event.
(Inherited from Control)

OnMove(EventArgs)
Raises the Move event.
(Inherited from Control)

OnNotifyMessage(Message)
Notifies the control of Windows messages.
(Inherited from Control)

OnPaddingChanged(EventArgs)
Raises the PaddingChanged event.
(Inherited from Control)

OnPaint(PaintEventArgs)
Raises the Paint event.
(Inherited from Control)

OnPaintBackground(PaintEventArgs)
Paints the background of the control.
(Inherited from Control)

OnParentBackColorChanged(EventArgs)
Raises the BackColorChanged event when the BackColor property value of the control's container changes.
(Inherited from Control)

OnParentBackgroundImageChanged(EventArgs)
Raises the BackgroundImageChanged event when the BackgroundImage property value of the control's container changes.
(Inherited from Control)

OnParentBindingContextChanged(EventArgs)
Raises the BindingContextChanged event when the BindingContext property value of the control's container changes.
(Inherited from Control)

OnParentChanged(EventArgs)
Raises the ParentChanged event.
(Inherited from Control)

OnParentCursorChanged(EventArgs)
Raises the CursorChanged event.
(Inherited from Control)

OnParentEnabledChanged(EventArgs)
Raises the EnabledChanged event when the Enabled property value of the control's container changes.
(Inherited from Control)

OnParentFontChanged(EventArgs)
Raises the FontChanged event when the Font property value of the control's container changes.
(Inherited from Control)

OnParentForeColorChanged(EventArgs)
Raises the ForeColorChanged event when the ForeColor property value of the control's container changes.
(Inherited from Control)

OnParentRightToLeftChanged(EventArgs)
Raises the RightToLeftChanged event when the RightToLeft property value of the control's container changes.
(Inherited from Control)

OnParentVisibleChanged(EventArgs)
Raises the VisibleChanged event when the Visible property value of the control's container changes.
(Inherited from Control)

OnPreviewKeyDown(PreviewKeyDownEventArgs)
Raises the PreviewKeyDown event.
(Inherited from Control)

OnPrint(PaintEventArgs)
Raises the Paint event.
(Inherited from Control)

OnQueryContinueDrag(QueryContinueDragEventArgs)
Raises the QueryContinueDrag event.
(Inherited from Control)

OnRegionChanged(EventArgs)
Raises the RegionChanged event.
(Inherited from Control)

OnResize(EventArgs)
Raises the Resize event.
(Inherited from Control)

OnRightToLeftChanged(EventArgs)
Raises the RightToLeftChanged event.
(Inherited from Control)

OnRightToLeftLayoutChanged(EventArgs)
Raises the RightToLeftLayout event.

OnSizeChanged(EventArgs)
Raises the SizeChanged event.
(Inherited from Control)

OnStyleChanged(EventArgs)
Raises the StyleChanged event.
(Inherited from Control)

OnSystemColorsChanged(EventArgs)
Raises the SystemColorsChanged event.
(Inherited from Control)

OnTabIndexChanged(EventArgs)
Raises the TabIndexChanged event.
(Inherited from Control)

OnTabStopChanged(EventArgs)
Raises the TabStopChanged event.
(Inherited from Control)

OnTextChanged(EventArgs)
Raises the TextChanged event.
(Inherited from Control)

OnValidated(EventArgs)
Raises the Validated event.
(Inherited from Control)

OnValidating(CancelEventArgs)
Raises the Validating event.
(Inherited from Control)

OnVisibleChanged(EventArgs)
Raises the VisibleChanged event.

(Inherited from Control)

PerformLayout()
Forces the control to apply layout logic to all its child controls.
(Inherited from Control)

PerformLayout(Control, String)
Forces the control to apply layout logic to all its child controls.
(Inherited from Control)

PerformStep()
Advances the current position of the progress bar by the amount of the Step property.

PointToClient(Point)
Computes the location of the specified screen point into client coordinates.
(Inherited from Control)

PointToScreen(Point)
Computes the location of the specified client point into screen coordinates.
(Inherited from Control)

PreProcessControlMessage(Message)
Preprocesses keyboard or input messages within the message loop before they are dispatched.
(Inherited from Control)

PreProcessMessage(Message)
Preprocesses keyboard or input messages within the message loop before they are dispatched.
(Inherited from Control)

ProcessCmdKey(Message, Keys)
Processes a command key.
(Inherited from Control)

ProcessDialogChar(Char)
Processes a dialog character.

(Inherited from Control)

ProcessDialogKey(Keys)
Processes a dialog key.
(Inherited from Control)

ProcessKeyEventArgs(Message)
Processes a key message and generates the appropriate control events.
(Inherited from Control)

ProcessKeyMessage(Message)
Processes a keyboard message.
(Inherited from Control)

ProcessKeyPreview(Message)
Previews a keyboard message.
(Inherited from Control)

ProcessMnemonic(Char)
Processes a mnemonic character.
(Inherited from Control)

RaiseDragEvent(Object, DragEventArgs)
Raises the appropriate drag event.
(Inherited from Control)

RaiseKeyEvent(Object, KeyEventArgs)
Raises the appropriate key event.
(Inherited from Control)

RaiseMouseEvent(Object, MouseEventArgs)
Raises the appropriate mouse event.
(Inherited from Control)

RaisePaintEvent(Object, PaintEventArgs)
Raises the appropriate paint event.
(Inherited from Control)

RecreateHandle()
Forces the re-creation of the handle for the control.
(Inherited from Control)

RectangleToClient(Rectangle)
Computes the size and location of the specified screen rectangle in client coordinates.
(Inherited from Control)

RectangleToScreen(Rectangle)
Computes the size and location of the specified client rectangle in screen coordinates.
(Inherited from Control)

Refresh()
Forces the control to invalidate its client area and immediately redraw itself and any child controls.
(Inherited from Control)

RescaleConstantsForDpi(Int32, Int32)
Provides constants for rescaling the control when a DPI change occurs.
(Inherited from Control)

ResetBackColor()
Resets the BackColor property to its default value.
(Inherited from Control)

ResetBindings()
Causes a control bound to the BindingSource to reread all the items in the list and refresh their displayed values.
(Inherited from Control)

ResetCursor()
Resets the Cursor property to its default value.
(Inherited from Control)

ResetFont()
Resets the Font property to its default value.

(Inherited from Control)

ResetForeColor()
Resets the ForeColor to its default value.

ResetImeMode()
Resets the ImeMode property to its default value.
(Inherited from Control)

ResetMouseEventArgs()
Resets the control to handle the MouseLeave event.
(Inherited from Control)

ResetRightToLeft()
Resets the RightToLeft property to its default value.
(Inherited from Control)

ResetText()
Resets the Text property to its default value (Empty).
(Inherited from Control)

ResumeLayout()
Resumes usual layout logic.
(Inherited from Control)

ResumeLayout(Boolean)
Resumes usual layout logic, optionally forcing an immediate layout of pending layout requests.
(Inherited from Control)

RtlTranslateAlignment(ContentAlignment)
Converts the specified ContentAlignment to the appropriate ContentAlignment to support right-to-left text.
(Inherited from Control)

RtlTranslateAlignment(HorizontalAlignment)
Converts the specified HorizontalAlignment to the appropriate HorizontalAlignment to support right-to-left text.

(Inherited from Control)

RtlTranslateAlignment(LeftRightAlignment)
Converts the specified LeftRightAlignment to the appropriate LeftRightAlignment to support right-to-left text.
(Inherited from Control)

RtlTranslateContent(ContentAlignment)
Converts the specified ContentAlignment to the appropriate ContentAlignment to support right-to-left text.
(Inherited from Control)

RtlTranslateHorizontal(HorizontalAlignment)
Converts the specified HorizontalAlignment to the appropriate HorizontalAlignment to support right-to-left text.
(Inherited from Control)

RtlTranslateLeftRight(LeftRightAlignment)
Converts the specified LeftRightAlignment to the appropriate LeftRightAlignment to support right-to-left text.
(Inherited from Control)

Scale(Single)
Obsolete.
Scales the control and any child controls.
(Inherited from Control)

Scale(Single, Single)
Obsolete.
Scales the entire control and any child controls.
(Inherited from Control)

Scale(SizeF)
Scales the control and all child controls by the specified scaling factor.
(Inherited from Control)

ScaleBitmapLogicalToDevice(Bitmap)
Scales a logical bitmap value to it's equivalent device unit value when a DPI

change occurs.
(Inherited from Control)

ScaleControl(SizeF, BoundsSpecified)
Scales a control's location, size, padding and margin.
(Inherited from Control)

ScaleCore(Single, Single)
This method is not relevant for this class.
(Inherited from Control)

Select()
Activates the control.
(Inherited from Control)

Select(Boolean, Boolean)
Activates a child control. Optionally specifies the direction in the tab order to select the control from.
(Inherited from Control)

SelectNextControl(Control, Boolean, Boolean, Boolean, Boolean)
Activates the next control.
(Inherited from Control)

SendToBack()
Sends the control to the back of the z-order.
(Inherited from Control)

SetAutoSizeMode(AutoSizeMode)
Sets a value indicating how a control will behave when its AutoSize property is enabled.
(Inherited from Control)

SetBounds(Int32, Int32, Int32, Int32)
Sets the bounds of the control to the specified location and size.
(Inherited from Control)

SetBounds(Int32, Int32, Int32, Int32, BoundsSpecified)

Sets the specified bounds of the control to the specified location and size.
(Inherited from Control)

SetBoundsCore(Int32, Int32, Int32, Int32, BoundsSpecified)
Performs the work of setting the specified bounds of this control.
(Inherited from Control)

SetClientSizeCore(Int32, Int32)
Sets the size of the client area of the control.
(Inherited from Control)

SetStyle(ControlStyles, Boolean)
Sets a specified ControlStyles flag to either true or false.
(Inherited from Control)

SetTopLevel(Boolean)
Sets the control as the top-level control.
(Inherited from Control)

SetVisibleCore(Boolean)
Sets the control to the specified visible state.
(Inherited from Control)

Show()
Displays the control to the user.
(Inherited from Control)

SizeFromClientSize(Size)
Determines the size of the entire control from the height and width of its client area.
(Inherited from Control)

SuspendLayout()
Temporarily suspends the layout logic for the control.
(Inherited from Control)

ToString()
Returns a string that represents the ProgressBar control.

Update()
Causes the control to redraw the invalidated regions within its client area.
(Inherited from Control)

UpdateBounds()
Updates the bounds of the control with the current size and location.
(Inherited from Control)

UpdateBounds(Int32, Int32, Int32, Int32)
Updates the bounds of the control with the specified size and location.
(Inherited from Control)

UpdateBounds(Int32, Int32, Int32, Int32, Int32, Int32)
Updates the bounds of the control with the specified size, location, and client size.
(Inherited from Control)

UpdateStyles()
Forces the assigned styles to be reapplied to the control.
(Inherited from Control)

UpdateZOrder()
Updates the control in its parent's z-order.
(Inherited from Control)

WndProc(Message)
Processes Windows messages.
(Inherited from Control)

Appendix C - Events

AutoSizeChanged
This event is not relevant for this class.
(Inherited from Control)

BackColorChanged
Occurs when the value of the BackColor property changes.
(Inherited from Control)

BackgroundImageChanged
Occurs when the value of the BackgroundImage property changes.

BackgroundImageLayoutChanged
Occurs when the value of the BackgroundImageLayout property changes.

BindingContextChanged
Occurs when the value of the BindingContext property changes.
(Inherited from Control)

CausesValidationChanged
Occurs when the value of the CausesValidation property changes.

ChangeUICues
Occurs when the focus or keyboard user interface (UI) cues change.
(Inherited from Control)

Click
Occurs when the control is clicked.
(Inherited from Control)

ClientSizeChanged
Occurs when the value of the ClientSize property changes.
(Inherited from Control)

ContextMenuChanged
Occurs when the value of the ContextMenu property changes.
(Inherited from Control)

ContextMenuStripChanged
Occurs when the value of the ContextMenuStrip property changes.
(Inherited from Control)

ControlAdded
Occurs when a new control is added to the Control.ControlCollection.
(Inherited from Control)

ControlRemoved
Occurs when a control is removed from the Control.ControlCollection.
(Inherited from Control)

CursorChanged
Occurs when the value of the Cursor property changes.
(Inherited from Control)

Disposed
Occurs when the component is disposed by a call to the Dispose() method.
(Inherited from Component)

DockChanged
Occurs when the value of the Dock property changes.
(Inherited from Control)

DoubleClick
Occurs when the user double-clicks the control.

DpiChangedAfterParent
Occurs when the DPI setting for a control is changed programmatically after the

DPI of its parent control or form has changed.
(Inherited from Control)

DpiChangedBeforeParent
Occurs when the DPI setting for a control is changed programmatically before a DPI change event for its parent control or form has occurred.
(Inherited from Control)

DragDrop
Occurs when a drag-and-drop operation is completed.
(Inherited from Control)

DragEnter
Occurs when an object is dragged into the control's bounds.
(Inherited from Control)

DragLeave
Occurs when an object is dragged out of the control's bounds.
(Inherited from Control)

DragOver
Occurs when an object is dragged over the control's bounds.
(Inherited from Control)

EnabledChanged
Occurs when the Enabled property value has changed.
(Inherited from Control)

Enter
Occurs when focus enters the ProgressBar control.

FontChanged
Occurs when the value of the Font property changes.

ForeColorChanged
Occurs when the ForeColor property value changes.
(Inherited from Control)

GiveFeedback
Occurs during a drag operation.
(Inherited from Control)

GotFocus
Occurs when the control receives focus.
(Inherited from Control)

HandleCreated
Occurs when a handle is created for the control.
(Inherited from Control)

HandleDestroyed
Occurs when the control's handle is in the process of being destroyed.
(Inherited from Control)

HelpRequested
Occurs when the user requests help for a control.
(Inherited from Control)

ImeModeChanged
Occurs when the value of the ImeMode property changes.

Invalidated
Occurs when a control's display requires redrawing.
(Inherited from Control)

KeyDown
Occurs when the user presses a key while the control has focus.

KeyPress
Occurs when the user presses a key while the control has focus.

KeyUp
Occurs when the user releases a key while the control has focus.

Layout
Occurs when a control should reposition its child controls.

(Inherited from Control)

Leave
Occurs when focus leaves the ProgressBar control.

LocationChanged
Occurs when the Location property value has changed.
(Inherited from Control)

LostFocus
Occurs when the control loses focus.
(Inherited from Control)

MarginChanged
Occurs when the control's margin changes.
(Inherited from Control)

MouseCaptureChanged
Occurs when the control loses mouse capture.
(Inherited from Control)

MouseClick
Occurs when the control is clicked by the mouse.
(Inherited from Control)

MouseDoubleClick
Occurs when the user double-clicks the control.

MouseDown
Occurs when the mouse pointer is over the control and a mouse button is pressed.
(Inherited from Control)

MouseEnter
Occurs when the mouse pointer enters the control.
(Inherited from Control)

MouseHover
Occurs when the mouse pointer rests on the control.

(Inherited from Control)

MouseLeave
Occurs when the mouse pointer leaves the control.
(Inherited from Control)

MouseMove
Occurs when the mouse pointer is moved over the control.
(Inherited from Control)

MouseUp
Occurs when the mouse pointer is over the control and a mouse button is released.
(Inherited from Control)

MouseWheel
Occurs when the mouse wheel moves while the control has focus.
(Inherited from Control)

Move
Occurs when the control is moved.
(Inherited from Control)

PaddingChanged
Occurs when the value of the Padding property changes.

Paint
Occurs when the ProgressBar is drawn.

ParentChanged
Occurs when the Parent property value changes.
(Inherited from Control)

PreviewKeyDown
Occurs before the KeyDown event when a key is pressed while focus is on this control.
(Inherited from Control)

QueryAccessibilityHelp

Occurs when AccessibleObject is providing help to accessibility applications.
(Inherited from Control)

QueryContinueDrag
Occurs during a drag-and-drop operation and enables the drag source to determine whether the drag-and-drop operation should be canceled.
(Inherited from Control)

RegionChanged
Occurs when the value of the Region property changes.
(Inherited from Control)

Resize
Occurs when the control is resized.
(Inherited from Control)

RightToLeftChanged
Occurs when the RightToLeft property value changes.
(Inherited from Control)

RightToLeftLayoutChanged
Occurs when the RightToLeftLayout property changes.

SizeChanged
Occurs when the Size property value changes.
(Inherited from Control)

StyleChanged
Occurs when the control style changes.
(Inherited from Control)

SystemColorsChanged
Occurs when the system colors change.
(Inherited from Control)

TabIndexChanged
Occurs when the TabIndex property value changes.
(Inherited from Control)

TabStopChanged
Occurs when the TabStop property changes.

TextChanged
Occurs when the Text property changes.

Validated
Occurs when the control is finished validating.
(Inherited from Control)

Validating
Occurs when the control is validating.
(Inherited from Control)

VisibleChanged
Occurs when the Visible property value changes.
(Inherited from Control)

Bibliography

[1] Bjarne Stroustrup. *The C++ Programming Language, 4th Edition 4th Edition.* Addison-Wesley Professional, 2013.

[2] Mark Lutz. *Programming Python: Powerful Object-Oriented Programming Fourth Edition.* O'Reilly, 2001.

[3] Mark Lutz. *Learning Python, 5th Edition.* O'Reilly, 2003.

[4] Mark J Price. *C# 9 and .NET 5 – Modern Cross-Platform Development: Build intelligent apps, websites, and services with Blazor, ASP.NET Core, and Entity Framework Core using Visual Studio Code, 5th Edition.* Packt Publishing, 2020.

[5] Adam Seebeck. *C# Fundamentals: C 9 and .NET 5 Paperback.* unQbd, 2020.

[6] Tom Archer Richard C Leinecker. *Visual C++ 6 Bible.* Wiley, 1998.

[7] John Lakos. *Large-Scale C++ Software Design.* Addison-Wesley Professional Computing Series, 1996.

[8] gRPC authors. *gRPC A high performance open source universal RPC framework.* `https://www.grpc.io`, 2021. World Wide Web.

Index

About The Author

Opubo G Benebo has been writin computer programs since his undergraduate engineering student, starting then with FORTRAN IV and FLOWTRAN and then through graduate school and grown into the developing codes in C/C++, Visual Basic, Visual C++ MFC, Java and some obsolete languages and platforms; and of course, currently an avid developer in python programming.

He has also taught code development in several computer programming school in the late 1990s ans early 2000s when he focus shifted to more entrepreneurial programming with a significant work on the development of software for the computational validation chemical and metabolic reactions utilizing electron-particle beam for ignition, as well as design and control of suchtechnologies as Air Pollutants Incinerators and Medical Diagnostic Machines.

He is a chemical engineer with B. Applied-Sc and MSc, and is also the founder of Induss SoftWerks, LLC, a software company with a dynamic growth-thrust. The founding of the software company was an overall entrepreneurial endeavour at founding a software company as one of a diversified synergistic conglomerate, for availing the group members with business support software.

Besides **PythonNet Programming** Contextual GUI Logical Designs for Beginners; he is also the author of other books including
Algae Cultivation Bioreactor Analysis, Okumaye Publishing Co, Inc 2018;
Batch Microbiological Reactors Designs; *Bioreactors Computational Designs Structures*, Okumaye Publishing Co, Inc 2015;
Designing Batch Fermentation Reactors; Design Basics, Okumaye Publishing Co, Inc 2013.

Generally, Opubo G Benebo may be reached through *opubo.benebo@nkassens.com* his email associated with the business operations through which he continues to undertake his entrepreneurial and founding corporations endeavours

Made in United States
Orlando, FL
04 November 2026

90201559R00150